FOREWORD

The International Institute of Seismology and Earthquake Engineering (IISEE) at Building Research Institute (BRI) was established in 1962 for the purpose of training young seismologists and earthquake engineers from earthquake-prone countries to help devote themselves to mitigation of earthquake disasters in their countries. As of March 2009, 1,380 participants from 96 countries and region have been trained since the start of the international training on seismology and earthquake engineering in 1960.

Now IISEE has three training courses; the Regular Course (Annual Training), the Individual Course and the Global Seismological Observation Course. Since the 2005-2006 Annual Training on Seismology and Earthquake Engineering, BRI, Japan International Cooperation Agency (JICA), and National Graduate Institute for Policy Studies (GRIPS) have cooperated to found the master course "Earthquake Disaster Mitigation Program" for getting "Master of Disaster Mitigation." Sixty-nine participants were awarded Master's Degrees till now. The Global Seismological Observation Course has started in 1995, designed to introduce up-to-date technologies and knowledge in the field of global seismological observation to the participants who are expected to play important roles in a global monitoring network on underground nuclear tests.

The Bulletin of the International Institute of Seismology and Earthquake Engineering is published annually as a medium of exchanging knowledge and techniques related to seismology and earthquake engineering in the world.

This issue of the Bulletin of IISEE contains twenty-five synopses of Master Papers by the Annual Training Course participants from eighteen countries.

I sincerely hope this Bulletin of IISEE contributes to dissemination and progress of disaster mitigation technology in earthquake-prone countries.

Nobuo Hurukawa
Director
International Institute of Seismology and
Earthquake Engineering

CONTENTS

Vol. 43

Synopses of Master Papers (the 2007-2008 course)

Synopses of Master Papers   *Bulletin of IISEE, 43, 1-6, 2009*

A SITE AMPLIFICATION STUDY USING OBSERVED RECORDS AT BMD, DHAKA FOR SEISMIC MICROZONATION

Md.Rubyet Kabir*   **Supervisor: Fumio Kaneko****
MEE07162

ABSTRACT

Dhaka city is currently seismically vulnerable both in natural and social condition, therefore seismic disaster management microzonation is indispensable. In this process, local site amplification plays an important role. For this purpose, due to shortage of soil data, a number of horizontal-to-vertical spectral ratio (H/V) for microtremor was already applied to estimate predominant period and amplification factor for geomorphological types, but without verification using real earthquake records. However, there are earthquake records obtained by the vertical array at the Bangladesh Meteorological Department (BMD), Dhaka.

In this study, in order to investigate an appropriate way to estimate site amplification due to subsurface soil for seismic microzonation of Dhaka, actual amplifications calculated using six actual small earthquake records are compared with the existing results of H/V for microtremor. This study provides the conclusions: (a) The predominant period estimated using H/V for micrtremor is in a good agreement with those estimated using the real small earthquake records, (b) The amplification factor of PGV and the amplitude spectra show some differences between the real small earthquake records and the existing results of H/V for microtremor, (c) the predominant periods of the existing H/V for microtremor can be used for microzonation of Dhaka city, whereas the amplification factors estimated from them cannot be used directly, but more careful investigations are required. Finally we expect the results of this paper can be a clue for preparation of practical seismic microzonation for Dhaka city.

Keywords: PGV Amplification, H/V Spectral Ratio, Microtremors, Microzonation, Predominant Period

1. INTRODUCTION

According to a report published by the United Nations IDNDR-RADIUS Initiative, Dhaka and Tehran are the cities at the highest earthquake disaster risk (Cardona et al., 1999). Dhaka mega city is overstressed due to its extremely high population, pressure on housing, transportation and other necessities for day by day life. Due to seismo-tectonic surrounding by having boundaries between Indian Plate and Eurasia Plate to the north, and between Indian Plate and Burma Plate to the east, Dhaka city has been experienced many historical earthquakes. Especially the 1897 Shillon Earthquakes (Ms=8.1) attacked Dhaka and provided severe damage (MMI 8). The existing study of seismic microzonation of Dhaka was composed of the probabilistic calculation of peak ground acceleration levels, and the estimation of predominant period of local amplification using H/V for microtremor and amplification factor of each geomorphological types (Kamal and Midorikawa, 2004;

---

*Bangladesh Meteorological Department
**OYO International Corporation

Kamal and Midorikawa, 2006). Though the applicability of H/V for microtremor has been said to have various questionable points (e. g., Satoh et al., 2001), it can reasonably be assumed that a major earthquake event in Dhaka region is capable of higher damage according to these results. But still the verification on site amplification characteristics by using actual earthquake records have not yet been realized, and necessary to certificate the accuracy. Fortunately, the Bangladesh Meteorological Department (BMD) has started seismic observation since August 2007, at Dhaka city especially with vertical array. This can provide a comparison between H/V for microtremor and actual earthquake motion amplification. This paper will also introduce an important clue for the practical seismic microzonation realization using existing data including H/V for microtremors, and more proper microtremor measurement.

2. Data

Six earthquakes were selected from one hundred three event files collected by the digital seismic network of BMD from August 31, 2007 to February 26. The existing microtremor data collected in Dhaka by Mr. M. Kamal were used also. Their observation points were near BMD and were employed by Kamal and Midorikawa (2004) and Kamal and Midorikawa (2006) for H/V analysis to estimate predominant period and amplification. They are labeled "MT" in the figures that appear hereafter.

3. ANALYSIS

3.1 Fourier Spectral Analysis

3.1.1 Fourier Transformation:
Firstly, the recorded waveform format is SEED. Then, Fourier spectra of the selected segments of two horizontal and the vertical components are calculated using Fast Fourier Transform (FFT) algorithm using Seismic Analysis Code (SAC). Then, before calculation of spectral ratios such as transfer function and H/V spectral ratios, smoothing technique using logarithmic window (Kon'no and Ohmachi, 1998) with a bandwidth coefficient b=15 is applied to reduce the distortion of peak amplitudes and to make it easy to show the peak of spectra.

3.1.2 Calculation of spectral ratios:
In order to calculate ratios of horizontal-to-horizontal components (that is the transfer function of horizontal components) and horizontal-to vertical components (H/V), the two horizontal components that are of North-South and East-West must be combined to a horizontal component. In this study, the square root average ($H=\sqrt{NS^2 + EW^2}$) and the geometric average ($H=\sqrt{NS \times EW}$) are used, they do not provide big difference in results.

3.2 Fourier Spectral Ratio

For calculating site amplification and making comparison with existing microtremor results with earthquake motion records, the velocity waveform data sets are divided into six time windows (Figure 1). Due to varying duration of P-S interval and the main part of S-wave, Case F) is more recommendable than others for the discussion on S-wave, Case D) rather than Case C) for discussion on S-coda. Considering on its stability, Fourier spectra of Case A) can be used as those of microtremor.

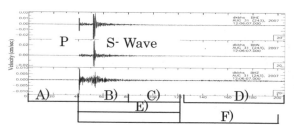

Figure 1. Division of velocity waveform of earthquake motion data. Case A) first 40sec before P-wave, Case B) 40.96 sec after P-wave, Case C) 40.96 sec after P-wave (2nd phase), Case D) last 81.92 sec, Case E) 81.92 sec after P-wave, and Case F) is 164 sec after P-wave.

3.3 Observed results of Spectral Ratios

3.3.1 Hs/Hb for earthquake:
Case F) shows clear three peaks at the periods 0.2, 0.4 and 1.1 second. The predominant one is given by the last one. The curve for microtremor (broken one) has peaks at almost same periods, but their amplitude at the peaks is smaller than those of earthquake records. Case B) and Case E) show similar shape to Case F). This suggests that the difference of Fourier spectra among Cases B), E) and F) is due to the source effect and cancelled out by calculating Hs/Hb. See Figure 2 (a).

3.3.2 Hs/Vs for earthquake and microtremor:
For Cases A), D), E) and F) a similarity to the curve for microtremor (broken one) is observed and for the latter two the more deviation of curves than the former two is observed. Hs/Vs is clearly different from Hs/Hb for all of Cases A), D), E) and F). For B) and Case C) the curves deviate much. See Figure 2 (b).

3.3.3 Hb/Vb for earthquake and microtremor:
For Cases A) and D), these are almost unity within factor of two in the period range shorter than 1sec, whereas in the longer period range these deviate more than the factor of two. For Case F), these can not fall in the factor of two around the unity. See Figure 2 (c).

3.3.4 Vs/Vb for earthquake and microtremor:
For all cases, this spectral ratio is not within factor of two around the unity in the period range shorter than 1 second, whereas in longer period range these for earthquakes (Case F)) show a good stability around the unity and these for noise (Case A)) show a slightly larger deviation but fall with in the factor of two. See Figure 2 (d).

3.3.5 Hs/Vs for microtremor versus Hs/Hb for earthquake
Above observation show that Hs/Vs for microtremor can not be an approximation of Hs/Hb for earthquake. Failure of the necessary conditions, i. e., Hb/Vb and Vs/Vb might be unity, suggest it too.

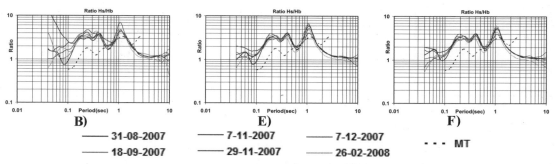

Figure 2 (a). Spectral Ratio for horizontal transfer function (Hs/Hb) for Case B), Case E) and Case F).

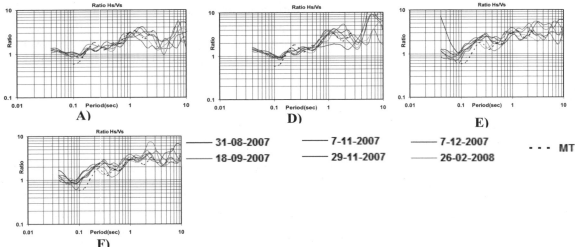

Figure 2 (b). Spectral Ratio for horizontal to vertical at surface (Hs/Vs) for Case A), Case D), Case E) and Case F).

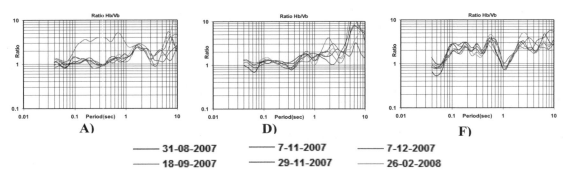

Figure 2 (c). Spectral Ratio for horizontal to vertical at borehole (Hb/Vb) for Case A), Case D), Case F).

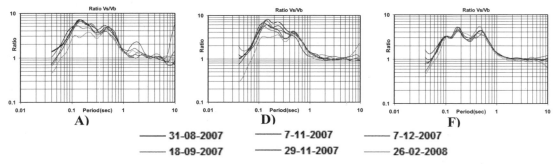

Figure 2 (d). Spectral Ratio for vertical transfer function (Vs/Vb) for Case A), Case D), and Case F).

3.4 PGV amplification and Peak Values of Spectral Ratio

Figure 3 (Left) shows the relation of the peak amplitude of Hs/Vs for S coda and for noise and Hs/Hb for earthquake. It can be confirmed that the peak amplitude of Hs/Vs for S coda and for noise have a correlation with PGV amplification better than those of Hs/Hb for earthquake. That of the existing H/V for microtremor results shows similar value to peak amplitude of Hs/Vs for S coda and for noise. Further, Figure 3 (Right) shows the same information as that of Figure 3 (Left) but the horizontal components are merged by geometrical average. It looks that the peak amplitudes of Hs/Hb for earthquake are similar as that shown in Figure 3 (Left), but the peak amplitudes of Hs/Vs for S coda become smaller and show a better correlation with PGV amplifications. In contrast, Hs/Vs for noise show a worse correlation in Figure 3 (Right). For both of those figures, Hs/Vs for S-coda and for noise almost fall within the factor of about 1.5 from PGV amplification.

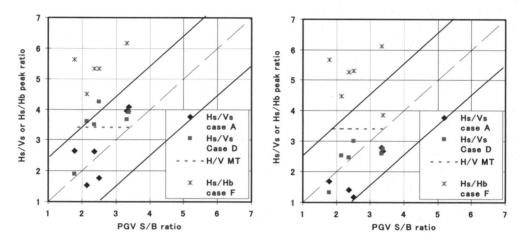

Figure 3. Relation of the existing H/V for microtremor results and the peak amplitude of Hs/Vs for S coda and for noise, Hs/Hb for earthquake with PGV amplification: horizontal components are merged (Left) by square root average and (Right) by geometric average

4. CONCLUSION

Based on the results of analysis using the earthquake records obtained at BMD, Dhaka, microtremor data of which duration is longer than 40 seconds with sampling frequency 100 Hz provide stable spectra for the period range from around 0.1 to 3 seconds. Thus, when targeting this period range, such duration is recommendable for getting stable spectra.

Predominant period estimated by H/V for microtremor is consistent with that of horizontal transfer function of earthquake motions Hs/Hb. This may be useful information for site characterization. Amplitude of H/V for microtremor at its predominant period shows a similarity with the peak amplitude of Hs/Vs of earthquake records, but significantly smaller than that of horizontal transfer function Hs/Hb.

The shapes of Fourier spectra as well as the spectral ratios except Hs/Vs are affected by the difference of wave contents between microtremor and earthquake motion, namely, the former is composed of mainly surface waves and the latter of body waves rather than surface waves. Even though shapes of H/V for microtremor and that of Hs/Vs of earthquake motions look similar, the former cannot be directly used in place of horizontal transfer function Hs/Hb. The basic assumption for Nakamura method is that Hb/Vb is unity or two and Vs/Vb is unity. Hb/Vb of microtremor becomes roughly unity perhaps because of surface waves' dominance, whereas that of earthquake motions becomes a little bit larger and complicated perhaps because of mixture of body waves. Vertical transfer functions Vs/Vb for both of microtremor and earthquake motions become unity in period range longer than 1 sec., whereas they have amplitudes from 3 to 5 in the shorter period range. Horizontal transfer function Hs/Hb for earthquake motions has clear peaks but that of microtremor show a broad peaks perhaps due to difference of wave contents mentioned above.

There are various problems to be solved for using H/V for microtremor as an estimate of horizontal transfer function Hs/Hb or PGV amplification factor. A conversion function if established, however, would enable us to estimate an amplification factor or spectra of amplification using H/V for microtremor. This suggests the necessity of conducting studies using much more vertical array observations for earthquake on applicability of seismograms and microtremor, and on effectiveness of non-linearity of soils during much stronger earthquake motion.

Even though several problems remain unsolved, this study introduces for the first time to Bangladesh the site characteristics during earthquake using actual earthquake records, by comparing microtremor with the actual earthquake motions.

5. RECOMMENDATION

For further studies it is necessary to collect more earthquake motions recorded at various sites including vertical arrays. The first priority, however, should be put on clarifying the soil structure, especially shear wave velocity structure, at BMD, Dhaka. The existing records may be useful to estimate the average velocity, but PS logging, seismic exploration or microtremor array measurement are highly recommended for further study.

Finally, this paper can give a clue for applicability of H/V for microtremor that is popularly conducted for practical seismic microzonation for Dhaka city, Bangladesh.

AKNOWLEDGEMENT

I would like to express deepest gratitude to Mr. Shukyo Segawa and Dr. Koichi Hasegawa, OYO International Corporation for his valuable guidance, discussions and generous help during the period of study. I am very grateful to all members of OYO International Corporation for their help and cooperation.

REFERENCES

Cardona, C., Davidson, R., and Villacis, C., 1999, *IDNDR*, Geohazard international.

Kamal, A. S. M. M. and Midorikawa, S., 2004, *International Journal of Applied Earth Observation and Geoinformation,* **6**, 111–125

Kamal, A. S. M. M., and Midorikawa, S., 2006, *IAEG 2006* paper number 457

Konno, K., and Ohmachi, T., 1998, *Bull. Seism. Soc. Am.* **88**, 1, pp. 228–24.

Nakamura, Y., 1989, *QR. RTRI,* **30**, 25-33.

Satoh, T., Kawase, H., and Matsushima, S., 2001, *Bull. Seism. Soc. Am.* **91**, 6, pp. 313–334.

Synopses of Master Papers   *Bulletin of IISEE, 43, 7-12, 2009*

MOMENT TENSOR AND HYPOCENTER DETERMINATION USING WAVEFORM DATA FROM THE NEW BANGLADESH SEISMIC NETWORK

Sayeed Ahmed Choudhury*   **Supervisor: Tatsuhiko HARA****
MEE07160

ABSTRACT

The Bangladesh Meteorological Department (BMD) has established a new digital seismic observation network in 2007. The data from this digital network have been used to determine a moment tensor using software developed by Dreger (2003). I analyzed the November 11, 2007 Bangladesh-India border earthquake. The obtained focal mechanism is consistent with that of the Global CMT catalog. This result suggests that it is possible for BMD to monitor and investigate the seismic activities by moment tensor determination procedure using data from the newly established Bangladesh Meteorological Department Seismic Network.

   I also performed hypocenter determination for the recent three earthquakes. The results for the two events that occurred in Bangladesh are consistent with the USGS hypocenters and felt reports. As for the event in Myanmar, the difference between their depths is large.

Keywords: Moment tensor inversion, Hypocenter.

INTRODUCTION

Seismicity

The north and east parts of Bangladesh is seismically active. The seismicity in these regions is attributed to the collision between the Indian plate and the Eurasian plate, and the subduction along the Indo-Myanmar range. Bangladesh has experienced large earthquakes in its history. Figure 1 shows historical and recent earthquakes in and around Bangladesh. The five historical and recent earthquakes among them occurred in Bangladesh, and the other five events occurred around Bangladesh. This kind of information can be used to evaluate possible damages caused by future earthquakes in and around Bangladesh.

New Digital Seismic Network of Bangladesh

In Bangladesh, the first seismic station was established in Chittagong in 1954. An analog seismometer was deployed at this station. Recognizing the increasing vulnerability for earthquakes, especially in large cities due to high density of population, unplanned infrastructure and their close proximity to India and Myanmar's seismically active areas, the GOB is implementing a project to establish three new seismic stations with upgrading the existing one for earthquake monitoring of in and around Bangladesh in 2007.

---

* Bangladesh Meteorological Department (BMD), Bangladesh.
**International Institute of Seismology and Earthquake Engineering, Tsukuba, Japan.

This nation-wide seismic network (shown in Fig. 2) consists of six digital broadband seismometers (two borehole and four vault type), two digital short-period seismometer, six digital triaxial accelerometers with GPS synchronization and five sets of GPS (Geodetic), etc. Networking among these four stations will be done by dedicated digital network for data transmission to the central data collection and processing center in the BMD headquarters in Dhaka.

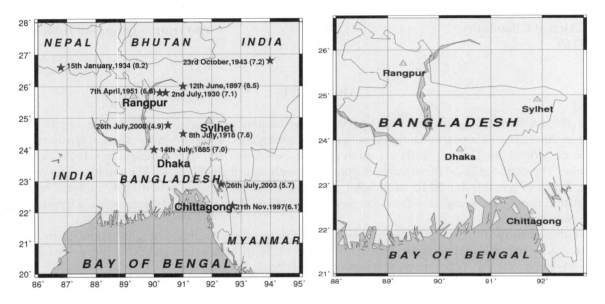

Figure 1. Some historical and recent earthquakes in and around Bangladesh.
(the data was obtained from the USGS).

Figure 2. The locations of the four stations of the Bangladesh seismic network.

TIME-DOMAIN MOMENT TENSOR PROCEDURE

Software

In this study, time-domain moment tensor inversion was carried out following Dreger and Romanowicz (1994) and Pasyanos *et al.*(1996). The software package of this technique has been used at the University of California, Berkeley Seismological Laboratory (BSL) since 1993 to automatically analyze events (ML>3.5) in Northern California. Also, this package has been implemented at the Japan National Institute for Earth Science and Disaster Prevention (NIED). This time-domain moment tensor inversion software package (TDMT_INV) is available at International Handbook of Earthquake and Engineering Seismology Handbook, part B. Green's functions were computed by using the FKRPROG software developed by Saikia (1994).

Method

The time-domain moment tensor inversion is performed as follows. First, waveform data are processed and prepared by the following steps: (i) baseline correction, (ii) deconvolution to obtain ground velocities, (iii) integration to calculate displacement, (iv) rotation to obtain transverse and radial components, (v) bandpass filtering, and (vi) re-sampling to 1 sps.

Then, Green's functions are calculated using the frequency wave number algorithm developed by Saikia (1994). The next step is to calculate time domain Green's function by the inverse FFT using the result in previous step. Then the bandpass filter with the same passband as was used for the observed data is applied.

Finally, linear least squares for deviatory moment tensor is preformed for a given source depth (both a spatial and temporal point source is assumed). The inversion yields the moment tensor which is decomposed into the scalar seismic moment, a double-couple moment tensor and a compensated linear vector dipole moment tensor. The decomposition is represented as percent double-couple and percent CLVD. The double-couple is further represented in terms of the strike, rake and dip of the two nodal planes. The basic methodology and the decomposition of the seismic moment tensor is described in Jost and Herrmann (1989).

The source depth is determined by finding the solution that yields the large variance reduction, (Dreger and Helmberger, 1993)

$$VR = \left[1 - \sum_i \sqrt{(data_i - synth_i)^2} \Big/ \sqrt{data_i^2} \right] * 100 \tag{1}$$

where *data* and *synth* are the data and Green's function time series, respectively, and the summation is performed for all the stations components.

Another measurement that is useful for determining source depth in regions where explosive events are unlikely is the RES/Pdc, the variance divided by the percent double-couple where,

$$RES / Pdc = \sum_i \sqrt{(data_i - synth_i)^2} / Pdc \tag{2}$$

Dividing the variance by the percent double-couple tends to deepen the minimum.

Crust Model

I selected a crust model from CRUST 2.0 (Bassin, Laske, and Masters, 2000) referring to the positon of the Chittagong station, and used this model for Green's function computation. CRUST 2.0 is a global crustal model specified on a 2x2 degree grid. Data were gathered from seismic experiments and averaged globally for similar geological and tectonic settings. The model and programs are available at http://mahi.ucsd.edu/Gabi/rem.dir/crust/crust2.html

Data

The event that occurred on the November 7, 2007 (the origin time: 07:10:21.72UTC after USGS) was selected for moment tensor inversion. The nearest station was Chittagong, and we used waveform data recorded at this station for the moment tensor inversion.

Table 1 shows the hypocenters and magnitudes issued by Bangladesh Meteorological Department, USGS, India Meteorological Department, and the Global CMT project (http://www.globalcmt.org/). The focal mechanism of this event in the Global CMT catalog is the strike skip fault.

Result

Figure 3 shows the result of the inversion for a focal depth of 15 km. We used the epicenter from the USGS in inversion, and the frequency band 0.03-0.04 Hz. The determining focal mechanism is a strike slip, which is consistent with the Global CMT solution.

We performed moment tensor inversions for depths of 5, 10, 15, 20, and 25 km, and compared variance reductions and variances divided by the corresponding percent double-couples. While the former is not sensitive to change of depths, the latter is more sensitive, and the focal depth of 15 km provides the smallest value.

Table 1. Hypocenters and magnitudes of the November 7, 2007 earthquake issued by Bangladesh Meteorological Department, USGS, India Meteorological Department, and the Global CMT project (* denotes lack of information).

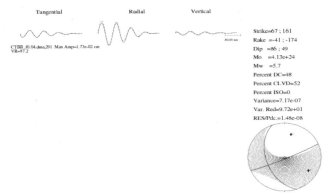

Organization	Lat (deg.)	Lon (deg.)	Depth (km)	M
USGS	22.15	92.39	28.7	5.5
BMD	22.14	92.38	*	6.0
IMS	22.1	92.5	15	5.3
Global CMT	22.15	92.5	25	5.5

Figure 3. The result of the moment tensor inversion (the focal depth is set to 15 km).

HYPOCENTER DETERMINATION

Data

I performed hypocenter determination to investigate whether it is possible to obtain reliable hypocenters using data from the new Bangladesh seismic network. I selected three events shown in Table 2. The epicenter of the latest event on the July 26, 2008 earthquake is at Mymenshing, and 120 km from the capacital city Dhaka. Its magnitude was 4.9. At least 25 people were injured at Dhaka.

I picked P and S arrival times of the Dhaka, Rangpur and Sylhet stations for the January 12, 2008 event (in the Bangladesh-India border region) and the December 7, 2007 event (in the Myanmar-India border region) using EDSP software. Bangladesh Meterological Department picked P and S arrival times of the Dhaka, Chittagong and Sylhet stations for the July 26, 2008 event (at Mymenshing, Bangladesh).

Table 2. The events selected for hypocenter determination (after the USGS).

Date	Time (UTC)	Lat (deg.)	Lon (deg.)	M	Depth (km)
26th Jul. 2008	18:51:48	24.79	90.51	4.9	1
12th Jan. 2008	22:44:47	22.76	92.33	5.0	33.8
7th Dec. 2007	6:56:34	23.50	94.57	4.8	113.3

Method and Crust Model

I used the program, HYPOCENTER, (Lienert et al., 1986) for hypocenter determination. This is an earthquake location method for locating local, regional and global earthquakes. Since the structures of the crust and upper mantle beneath Bangladesh is not precisely studied, I used model iasp91 (Kennett and Engdahl, 1991) as an earth model in this study. The initial depth guess is set to 0 km.

Result

Table 3 shows the determined hypocenters, and Figure 4 shows comparison between these epicenters and those from the USGS. The results for the events on the July 26, 2008 and the January 12, 2008 are consistent with the USGS hypocenters. As for the former event, the obtained epicenter is also consistent with felt repots. As for the event on the December 7, 2007, which occurred in Myanmar, although the epicenter of this study is close to that of the USGS, the difference between the depths is large. This result suggests the difficulty to locate events outside the network.

In all the cases, data from the three stations are available, and data from one station is missing. It is desirable that data from all of the stations are continuously available.

Table 3. The hypocenters determined in this study.

Date	Lat (deg.)	Lon (deg.)	Depth (km)
26th Jul. 2008	24.98	90.47	10.8
12th Jan. 2008	22.66	92.17	10.0
7th Dec. 2007	23.42	94.24	10.0

Figure 4. Comparison between the epicenters determin study (red stars) and those from the USGS (blue stars).

CONCLUSION

In this study, I applied the moment tensor inversion code developed by Dreger to the waveform data from the new broadband seismic network in Bangladesh. I also performed hypocenter determination of the recent three earthquakes using data from this network.

The focal mechanism solution obtained in this study is consistent with the Global CMT solution for the same event. The results of hypocenter determination for the events in Bangladesh are consistent with hypocenters determined by the USGS.

These results suggest that it will be possible to use both programs for further investigations of events in Bangladesh. Based on the achievements in this study, I am going to perform moment tensor inversion using the procedure of this study. I also plan to relocate previous events using the program HYPOCENTER.

RECOMMENDATIONS

Accurate hypocenter is necessary to perform moment tensor inversion used in this study. At present, there are four stations running in the Bangladesh seismic network. Due to communication problems and unstable power supplies, data from all the stations are not always available. It is desirable to increase the number of stations to enhance the capability of hypocenter determination. Also, data exchange among neighboring countries will be effective to improve accuracy of determination of earthquake source parameters.

To improve moment tensor determination, accurate crust model is necessary. Analyses using phase data and waveform data should be carried out to construct an appropriate crust model for earthquake monitoring. Through these improvements, I believe that our network will successfully work for earthquake monitoring in my country in the near future.

ACKNOWLEDGEMENTS

I would like to thank Dr. Wang Hongti Prossor, Institute of Earthquake Predication, China Earthquake Administration who supported my study by giving information about data and Instrument of our network.

REFERENCES

Bassin, C., Laske, G., and Masters, G., 2000, EOS Trans AGU, 81, F897.
Dreger, D.S., 2003, International Handbook of Earthquake and Engineering Seismology, W.H.K.Lee, Kanamori, H., Jennings, P.C., and Kisslinger C. (Eds.), Academic Press.
Dreger, D.S. and Helmberger, D.V., 1993, J. Geophys. Res., 98, 8107-8125.
Dreger, D.S. and Romanowicz, B., 1994, U. S. Geol. Surv. Open-file Rept., 94-176, 301-309.
Jost, M.L. and Herrmann, R., 1989, Seis. Res. Lett., 60, 37-57.
Lienert, B.R., Berg, E., and Frazer, L.N., 1986, Bull. Seism. Soc. Am., 76, 771-783.
Pasyanos, M, Dreger, D.S., and Romanowicz, B., 1996, Bull. Seism. Soc. Am., 86, 1255-1269.
Saikia, C.K., 1994. Geophys. J. Int., 118, 142-158.
Web site: Global Crustal Model is available here http://mahi.ucsd.edu/Gabi/rem.dir/crust/crust2.html

Synopsis of Master Papers   *Bulletin of IISEE, **43**, 13-18, 2009*

MOMENT TENSOR AND SOURCE PROCESS OF EARTHQUAKES IN FIJI REGION OBTAINED BY WAVEFORM INVERSION

Seru Sefanaia*   **Supervisor:**   **Yuji YAGI****
MEE07167

ABSTRACT

We evaluated the quality of the local seismic waveform recorded at Yasawa station established by the Mineral Resources Department (MRD, Fiji), and tried to estimate moment tensor solution by using this waveform data. We also estimated moment tensor solution and seismic source process by using teleseismic body wave so as to investigate the faulting system in Fiji region. As for the evaluation of the quality of the waveform recorded at the Yasawa station, we found that EW and NS components were down frequently, and moment tensor solution obtained by 3 components of the Yasawa station is totally opposite to the Harvard CMT solution. This result may suggest that the polarity of the local seismometer components was in a reverse direction and magnification of the seismometer response information is not correct at this stage. We obtained a consistant result when we divided by -10 to observed waveform. Moment tensor inversion analysis was also carried out for 30 events by using teleseismic body-wave downloaded from the IRIS-DMC. Final results are well consistent to Harvard CMT solution. Rupture source process analysis was conducted for the two deep intraslab events that occurred respectively along the Tonga Kermedec subduction zone and two shallow depth events which occurred respectively along the transform fault zones. The rupture processes for the two deep earthquakes are characterized by the rupture propagating mainly along the strike of the fault plane. The two shallow events occurred along the transition zone also characterized by the rupture propagating along the strike of the Fiji Fracture Zone but are controlled by the geometry of the fault.

Keywords: Moment tensor inversion, Rupture process inversion, teleseismic data, near-source data.

INTRODUCTION

Earthquake focal mechanisms are basic and important information in seismology and have been utilized for understanding regional tectonic stress fields, source mechanisms of large earthquakes, simulation of strong motion, faulting systems, tsunami generation and so on. It is important to estimate focal mechanisms of earthquakes routinely as well as to make a homogenous catalogue of moment tensors (or focal mechanisms) from small to large earthquakes for future references. Moment tensor and rupture process inversion analysis has not so far being carried out in Fiji. Therefore in this study, we evaluated the quality of the local seismic waveform data obtained from the Yasawa broadband station established by Mineral Resources Department (MRD, Fiji), then we tried to estimate moment tensor solution by using this waveform data and as well as to estimate moment tensor solution and seismic source process using teleseismic body wave so as to investigate the faulting system in the Fiji region.

---

*Mineral Resources Department (Seismology Section) of Fiji.
**Ass. Professor, Institute of Physical Sciences, University of Tsukuba, Japan.

We used both near-source data recorded at Yasawa broadband station established by Mineral Resources Department, Fiji with range from March 2003 to December 2005 as well as teleseismic body-wave (P-wave) data recorded at FDSN and GSN network stations collected by the Data Management Center of the Incorporated Research Institution for Seismology (IRIS-DMC).

Table 1: Shows the quality of Yasawa station components

Months	Year 2003 Components			Year 2004 Components			Year 2005 Components		
	UD	EW	NS	UD	EW	NS	UD	EW	NS
Jan	not yet being established			o	x	o	o	o	o
Feb				o	o	x	o	o	o
Mar	o	x	o	o	x	x	o	x	o
Apr	no data			o	o	x	o	o	o
May	o	x	o	o	x	x	o	o	o
June	o	x	o	o	o	o	o	o	o
July	no data			no data			o	o	o
Aug	no data			no data			o	o	o
Sept	o	x	o	o	o	o	o	o	o
Oct	o	x	x	o	o	x	o	o	o
Nov	o	o	o	x	o	o	o	o	o
Dec	o	x	o	o	o	o	o	o	o

o – Waveforms well recorded.
x – Component broken.

The raw continuous Yasawa waveform data was converted from nanometrics to seed format by makeseed software provided by Nanometrics Inc. Further, we obtained the waveform data in SAC format by using rdseed software which was available from the IRIS-DMC homepage. The broadband data were converted and numerically integrated from volts into the ground velocity cm/s and then band-passed between 0.01 and 0.1 Hz to reduce the effect of fine 3D structure and detailed source process. Table 1 shows the quality of waveforms recorded at the Yasawa station in each period. For teleseismic body-waves we retrieved and selected the IRIS-DMC stations from the viewpoint of good azimuthal coverage with the distance range from 30° to 90°. We referred to earthquake event list from the Harvard CMT catalogue from year 1990 to present. Thirty events with Mw > 6.5 occurred within Fiji region were analyzed for moment tensor inversion and 4 of these events were analyzed for their source rupture processes.

THEORY AND METHODOLOGY

Moment Tensor Inversion

To obtain moment tensor components of overall earthquake, we represented the seismic source process as point source model.

$$u_j(t) = \sum_{q=1}^{5} G_{jq}(t,\xi_c) * M_q'(t,\xi_c) + e_o + e_m$$

$$= \sum_{q=1}^{5} M_q'' \big(G_{jq}(t,\xi_c) * T(t) \big) + e_o + e_m \tag{1}$$

where M_q' and M_q'' are moment tensor at centroid of source ξ_c, $T(t)$ is source time function, and e_m is modeling error. For simplicity, the observation equation of (1) can be rewritten in vector form:

$$\mathbf{d}_j = \mathbf{H}(T(t),\xi_c)_j \mathbf{a} + \mathbf{e}_j \tag{2}$$

where **d** and e are N-dimensional data and error vectors, respectively, **a** is a 5-dimensional model parameter vector, **H** is a N x 5 coefficient matrix. The solution of the above matrix equation is obtained by using least square approach, if we assumed source time function and the centroid location. We determined the depth of centroid and the duration and shape of source time function by grid search method since these are needed for moment tensor inversion for the analysis of teleseismic body wave. For analyses of near-source data, we applied low pass filter to mitigate source time function effect and then estimated the depth of centroid by grid search method.

Rupture Process Inversion

To estimate rupture process, we basically followed the formulation by Yagi and Fukahata (2008). A vertical component of P-wave at a station j due to a shear dislocation source on a fault plane S is given by

$$u_j(t) = \sum_{q=1}^{2} \iint G_{jq}(t,\xi) * \dot{D}_q'(t,\xi) dS + e_o, \tag{3}$$

where G_{jq} is a Green's function, $\dot{D}_q'$ is a spatio-temporal slip-rate distribution, and $*$ denotes the convolution operator in time domain. To formulate the inverse problem, we represent the spatio-temporal slip-rate distribution $\dot{D}_q'$ by linear combination of a finite number of basis functions:

$$\dot{D}_q'(t,\xi_c) = \sum_{k=1}^{K} \sum_{l=1}^{L} a_{klq} X_k(\xi) T(t) + \delta\dot{D}_q'(t,\xi), \tag{4}$$

where a_{klq} are model parameters to be determined from observed data. $X_k(\xi)$ and $T_l(t)$ are the basis functions for space and time, respectively. We can rewrite in vector form:

$$\mathbf{d}_j = \mathbf{H}_j \mathbf{a} + \mathbf{e}_{mj}, \tag{5}$$

where $\mathbf{d}_j$ and $\mathbf{e}_{mj}$ are N_j-dimensional data and error vectors, respectively, **a** is a M-dimensional model parameter vector, and $\mathbf{H}_j$ is a $N_j \times M$ coefficient matrix.

RESULTS AND DISCUSSIONS

The results of this study are divided into three (3) main parts:

Part I: Moment Tensor Inversion using near source data obtained by Local Seismic Station

The local data set obtained for this study was from the period of its establishment in March, 2003 to December, 2005. We estimated the centroid moment tensor solution by using local seismic data which contains three components waveform. Green's function for near-field was calculated using program by Kohketsu (1985), and performed waveform inversion using program by Yagi (2006). To obtain the location of centroid, we applied the grid search method.

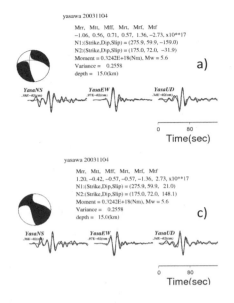

b)

a)

c)

Figure 1. Moment tensor solutions. a) Result obtained by this study using local seismic data recorded at Yasawa station. b) Harvard CMT solution. c) Result for the reverse polarity waveform. The red curve is the theoretical waveform and the black curve is the observed waveform.

We found the inconsistencies between this study solution (Figure 1a) and the Harvard CMT solution even if waveform fitting is good (Figure 1b). The result for this study is totally opposite to the Harvard CMT solution. One of the possibilities for this difference may be the polarity set up (EW and NS) components is in a reverse direction. If we divide by -1 to the waveform in our SAC program, the focal mechanism is well consistent with the result of Harvard CMT (Figure 1c). We should point out that our result (Figure 1c) is ten times larger in seismic moment than the Harvard result. This result implies that the magnification of the Yasawa seismometer information is not correct at this present time.

Part II: Moment Tensor Inversion by using Teleseismic Body Waveforms.

In Part II of this study, 30 events (Mw >6.5) that occurred within the Fiji region were analyzed by using teleseismic body waveforms downloaded from the Data Management Center of the Incorporated Research Institutions for Seismology (IRIS-DMC). We performed moment tensor inversion analysis for these local events by using moment tensor inversion programming coded by Yagi (2008). The depth of hypocenter, the seismic moment (Mo) and the final focal mechanism 'beachball' solutions of this study was computed with that of Harvard GMT solutions.

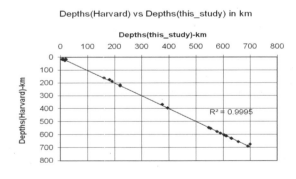

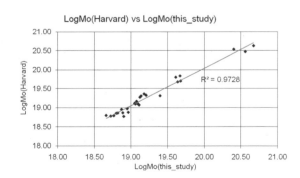

Figure 2. Graph of Depths (Harvard Solution)-km vs. Depths (this_study) in km.

Figure 3. Graph of LogMo (Harvard Solution) vs. LogMo (this_study)

Our results reveal the consistencies with the depths of hypocenters (Figure 2) and logMo (Figure 3) in the two solutions. Such a difference observed is due to that the Harvard University routinely carried out the overall waveform picking in their moment tensor analysis, whereas in this study utilizes only the P-wave for such analysis of moment tensor.

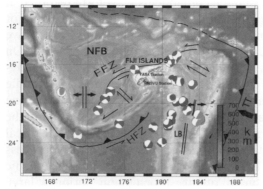

Figure 4 exhibits the final moment tensor solution of this_study plot as "beachballs" on the Fiji region map. Shallow depth events with strike-slip faulting type occurred mainly along the two Fracture zones, the Fiji Fracture Zone and the Hunter Fracture Zone. The rest are intermediate to deep intraslab earthquakes along the two subduction zones with mostly reverse and normal faulting type.

Figure 4. Focal mechanism plot of this study.

Part III: Analysis of rupture source process

Four earthquakes were analyzed for their rupture source processes during their occurrences. Two deep intraslab events which occurred along the Tonga subduction zone and two events were of shallow depths occurred along the Fiji Fracture Zone. We applied inversion code originally given by Yagi and Fukahata (2008) so that the constraints of smoothness and positivity were imposed on the solution used in this analysis. To detect actual fault plane, we found the fault plane and average slip direction which show best fitting using grid search method.

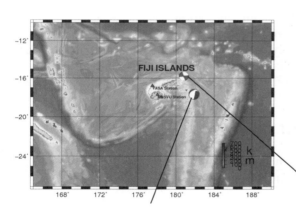

In general, for deep earthquakes rupture propagates mainly along the strike direction due to the thinness of the seismogenic zones at deeper depths as proved by our results shown on Figure 6. As for shallow earthquakes along the Fiji Fracture Zone, the rupture propagates along the strike (NE and SW) direction and breaks some part of the fault, however the propagation of the rupture seems to be controlled by the geometry of the fault as proved by this study solution shown on Figure 7.

Figure 5. Indicates the location of the two events plotted as "beachballs" on Fiji map

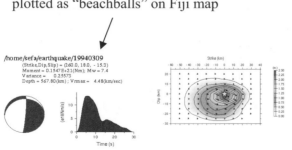 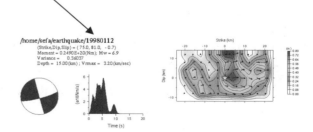

Figure 6. Shows the final results of inversion for one deep event 19940309. The star indicates the location of the initial break.

Figure 7. Shows the final results of inversion for one of the shallow depth earthquake 19980112. The star indicates the location of the initial break.

CONCLUSION

In this study, we first investigate the quality of waveforms recorded at Yasawa station which is operated by the Mineral Resources Department of Fiji. Using three components waveform recorded at Yasawa, we tried to estimate the centroid moment tensor solution so as to compare it with the result of Harvard CMT. We found that the solution by this study is not consistent with the Harvard CMT. If we divided by -10 to the observed waveform, we can obtain a consistent result as the Harvard CMT. On the other hand, moment tensor inversion analysis was carried out by using teleseismic body-waveforms for 30 past local earthquakes (Mw>6.5) retrieved from IRIS-DMC. We found that middle earthquake with shallow depth located along the two major fracture zones, the Fiji Fracture Zone (FFZ) and the Hunter Fracture Zone (HFZ) and intermediate and deep earthquakes located within the two subducting slab, the Tonga Kermedec subduction zone and the Vanuatu Trench. The focal mechanisms are well consistent with tectonic setting. In the final part of this study, source process inversion was performed for four events. As for deep earthquake, since thickness of seismogenic zones within the slab decreases with depths, the direction of rupture propagation during a large earthquake should be mainly along the strike of the slab. For shallow earthquake occurred within the Fiji Fracture Zone, rupture propagated along strike direction, NE and SW direction or parallel to the direction of the Fiji Fracture Zone. In general, for shallow earthquakes during such rupture, it breaks a part of the fault and the rupture propagation direction seems to be controlled by the orientation or the geometry of the fracture zone. Our results agree well to the theory.

RECOMMENDATION

Based upon the purpose of this study, the following recommendations are put forward.

1. To investigate and check the polarity of Yasawa broadband station as well as the magnification of the seismometer information and efforts should be made for our Mineral Resources Department (MRD) to monitor and maintain this station.
2. Also to check and maintain our CTBTO broadband station.
3. Moment tensor analysis to be carried out routinely for all medium-large size earthquakes that will occur within the Fiji region by using both near source data and teleseismic body-wave.
4. Database for this analysis should be compiled and accessed as references for future purposes.

ACKNOWLEDGEMENT

I would like to take this great opportunity to express my heartfelt thanks to my advisor Dr Hiroshi Inoue for the important advice given all the time especially regarding our seismic network in Fiji.

REFERENCES

Dregar, D., Ritsema, J. and Pasyanos, M., 1995, Geophys. Res. Lett., Vol. 22, 8, 997-1000.

Kohketsu, K., 1985, J. Phys. Earth, 33, 121-131.

Rognvaldsson, S. T. and Slunga, R., 1993, Bull Seis. Soc. Am. Vol. 83, 4, 1232-1247.

Tibi, R., Estabrook, C. H. and Bock G., 1999, Geophys. J. Int. (1999), 138, 625-642.

Yagi, Y., Kikuchi, M., 2000, Geophys. Res. Lett., Vol. 27, 13, 1969-1972.

Yagi, Y., Mikumo, T., Pacheco, J. and Reyes, G., 2004, Bull Seis. Soc. Am. Vol. 94, 5, 1795-1807.

Yagi, Y., 2007-2008, IISEE/BRI.

RELOCATION OF THE MACHAZE AND LACERDA EARTHQUAKES IN MOZAMBIQUE AND THE RUPTURE PROCESS OF THE 2006 Mw7.0 MACHAZE EARTHQUAKE

Paulino C. FEITIO*   Supervisors: Nobuo HURUKAWA**
MEE07165   Toshiaki YOKOI**

ABSTRACT

In the present study, the 2006 Machaze earthquake (Mw7.0) and the 2006 Lacerda earthquake (Mw5.6) in Mozambique are relocated. The main purpose of the relocation is to determine fault planes related to those earthquakes. The result of the relocation of the Machaze earthquake indicates that the strike and dip of the fault plane is about 172° and 65° westward, respectively, and it represents a normal faulting. Analysis of the Lacerda cluster suggests that this is an earthquake swarm. The fault plane corresponding to the biggest event presents a strike and dip of about 350° and 40°, respectively, dipping to the east. The rupture process of the Machaze earthquake is also determined in the present study. The result of the slip inversion shows that two asperities characterize the Machaze earthquake. The maximum slip is 3.4 m located in the south asperity near the initial break point and the asperity in the north of the initial break point has a slip of about 2.5 m. The last asperity is located at surface and generated the most prominent offset observed at the surface. The aftershocks are located near the two asperities. Both analyses show a size of the fault plane corresponding to the mainshock about 50 km in length.

Key words: Mozambique earthquake, relocation, joint hypocenter determination, slip inversion.

INTRODUCTION

African continent is divided in two major tectonic blocks, Nubia and Somalia. The boundary between these blocks is along the East African Rift System (EARS). Mozambique is located in the boundary zone between Nubian plate and Somalia plate, at the southern end of the EARS. The seismicity in the eastern Africa is related to the tectonics of the region.

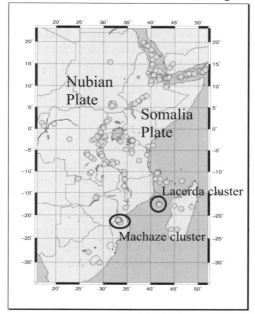

Figure 1. The seismicity map of eastern Africa region. The data used is from from Global CMT within the range from 1976 to 2007

The majority of events are located along the major tectonic lines, mainly along the EARS (Figure 1). The number of earthquakes occurring in the region, according to the historical recordings including the last strongest (Mw7.0) on February 22, 2006 at 22:19 UTC, with epicenter in Machaze, in the western province of Manica, denotes that earthquakes must be considered as one of the geohazards in the country. In the same year, on September 24, one event with magnitude Mw5.6 occurred in the offshore of Mozambique Channel. Both Machaze and Lacerda events are located in the south end of western and eastern branches of the EARS, respectively (circles in Figure 1). Precise earthquake locations play an important role in understanding earthquake source process and one of the most important parameters among the seismological ones on that strong motion estimation. The strong motion prediction is one of the key for earthquake disaster mitigation.

*National Directorate of Geology (DNG), Mozambique.
**International Institute of Seismology and Earthquake Engineering, Japan

In the present study the Machaze and Lacerda clusters are simultaneously relocated to determine the fault plane and seismic plane, respectively. Moment tensor inversion and Rupture process of the Machaze event are also determined in this study using the inversion technique developed by Yagi and Fukahata (2008).

METHODS

Three methods are used in the present study. The modified joint hypocenter determination (MJHD) method (Hurukawa and Imoto 1990, 1992) is used to determine the fault plane for the Machaze and the Lacerda earthquakes and two inversion methods are used to determine the moment tensor components and the rupture model for the Machaze earthquake, respectively. The MJHD technique is designed to minimize all travel-time residuals simultaneously and to find a common set of station corrections. This method removes the effect of horizontal heterogeneity of the earth through the introduction of station corrections. Due to the trade-off between station corrections and focal depth of earthquakes, the original JHD method becomes unstable and unreliable. The MJHD method introduced by Hurukawa and Imoto (1990, 1992) and Hurukawa (1995, 1998) overcome this problem by introducing the following priors. The station correction is independent of the distance and the azimuth from the center of the region to the station. This is very effective, especially in the case in which no earthquakes are observed clearly at all stations.

In order to understand the fault rupture process, waveform modeling is one of the most powerful tools available. The method developed by Yagi and Fukahata (2008) is used, in which covariance components in analysis of densely sampled observed data are considered. The source depth and source time function are determined using a grid search method. Assuming that those parameters are known, the moment tensor components were determined by inversion technique. For the slip inversion, the seismic fault is gridded into many sub-faults. Then the whole rupture process is considered to be the summation of all these sub-events, which are considered point source, respectively. The unknown parameters include the amount, direction and temporal change of slip at each grid point. The result of the inversion which gives the amount of slip at each grid point is obtained using least square fitting. From this inversion the initial depth can be corrected, hypocenter can be relatively accurate located, the initial rupture time, and the moment tensor can be obtained.

DATA AND RESULTS OF RELOCATION

The data used for relocation of both Machaze cluster and Lacerda swarm consist of P-arrival times from International Seismological Centre (ISC) and National Earthquake Information Center (NEIC) operated by United States Geological Survey (USGS). From ISC, the data was retrieved within the period between January 1, 1964 and December 31, 2004 and from NEIC the data ranges between January 1, 2005 and December 31, 2007.

Machaze cluster

The criterion used for station and events selection is MSTN (minimum number of stations which record the event) is 15 and for the MEVN (minimum number of events recorded by each station) is also 15. By using this criterion initially 69 stations were selected worldwide to locate 103 events. Figure 2 shows the hypocenters distribution determined by ISC and NEIC-USGS. The fault mechanism used is obtained from Global CMT catalog. Since depths of hypocenters were fixed at 10 km, it is difficult to identify the fault plane of the mainshock. The MJHD method is used to relocate the Machaze event. In these analyses only hypocenters with travel time residuals less than 2.0 seconds are relocated.

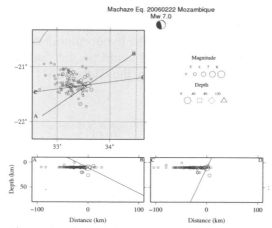

Figure 2. The epicentral distribution and two vertical cross sections along A-B and C-D lines, which are perpendicular to the strikes of the two nodal planes of the Glonbal CMT, are illustrated. The size of symbols represents the magnitude of the event and the shapes and the colors of the symbols denote the depth.

Observing the cross sections in Figure 3, it seems to be clear that the nodal plane shown in the cross section C-D is the fault plane of the Machaze earthquake. Figure 4 shows only immediate aftershoks.

Although there is some scatter in hypocenter distribution in horizontal direction, it still clear that the nodal plane shown in the cross section C-D in Figure 4 is the fault plane of the Machaze event. The strike and dip of this fault are 172° and 65°, respectively. The length of fault plane is estimated to be approximately 50 km and the epicenter is located near the center of the fault. This event has only one immediate foreshock observed in the relocated results and the epicenter is located less than 1 km from the mainshock. The time difference between the foreshock and mainshock is approximately 27 hours and 8 minutes.

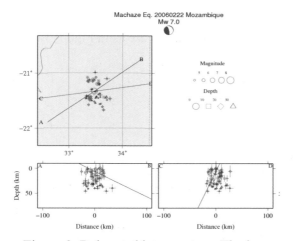

Figure 3. Relocated hypocenters. The bars in the circles represent the standard deviation. The sym,bols are the same as in Figure 2

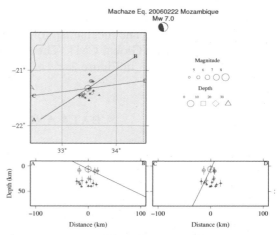

Figure 4. Immediate aftershocks within a week after the mainshock of the Machaze earthquake. The symbols are the same as in Figure 3.

Lacerda cluster

For the case of the Lacerda cluster the criteria used is as follows: MSTN is 12 and MEVN is also 12. In this relocation 21 stations were used to locate 15 events. Figure 5 shows the epicentral distribution determined using data from NEIC-USGS. Analyses of the relocated events shows that this cluster is composed by events with magnitude ranging between Mw4.0 and Mw5.6 and the biggest event (Mw5.6) is one of the last in the cluster, which suggest that the cluster can be considered an earthquake swarm. The fault mecanism used is obtained from Global CMT catalog, considering the biggest event.

The focal depth is fixed at 10 km, it is impossible to identify the fault plane. Figure 6 shows the results of relocation. It is clear that all events follow the N-S lineament, which is consistent with the tectonic lineament of the Lacerda graben. Although the scatter is large, the hypocenter distribution of these events suggests that the nodal plane shown in cross section A-B could be the fault plane related to the major event. The strike and dip of the fault plane determined by the Global CMT is 350° and 40°, respectively. This fault plane dips to the east.

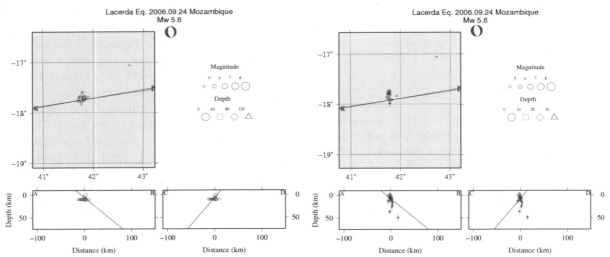

Figure 5. Epicentral distribution map using data from NEIC-USGS. The mechanism of the event considered is from Global CMT. The symbols are as in Figure 2.

Figure 6. Relocated events from Lacerda cluster. The symbols are as in Figure 3.

DATA AND RESULT OF SLIP INVERSION

The rupture model of Machaze earthquake is determined in the present study using the inversion method by Yagi and Fukahata (2008). The data used was retrieved from Data Management Center (DMC) of Incorporated Research Institute of Seismology (IRIS). Fifteen vertical components from broadband seismograph stations are used to pick P-arrivals. To retrieve the source information the stations were selected within the epicentral distance between 30° and 90°.

For slip distribution waveforms are inverted for both nodal planes. For the nodal plane I, 13x10 grid points with spacing of 4 km was adopted. For the nodal plane II, grid points with 13x13 and spacing of 4 km was adopted. The fault models are determined for both nodal planes. The hypocenter is determined by grid search method varying the depth for every 2.0 km for both fault models. According to Figure 7, the differences between the variances of the strike, dip and rake of both fault models are small, but the fault model I present slightly smaller variances. Using the result from aftershock distribution it is assumed that fault model I is the actual fault plane. The spatio-temporal slip distribution is presented in the Figure 8d.

Fault Model 1

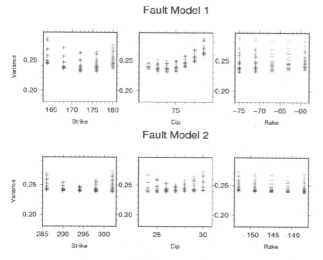

Fault Model 2

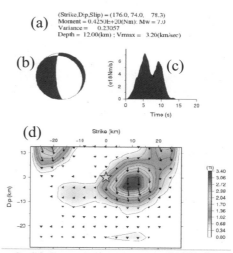

Figure 7. Comparison of the variances obtained for fault model I and fault model II.

Figure 8. (a) Source parameters, (b) fault mode I mechanism of the event, (c) source time function, (d) amount of slip in each grid point. The arrows indicate the slip direction of the hangingwall.

The final source parameters of the Machaze earthquake determined in the present study can be summarized as follows: Strike, dip and rake, 176, 74 and -78.3, respectively; the initial break point is located at 12.0 km depth; the source duration is 13 seconds (Figure 8c); there are two asperities, the maximum slip is 3.4 m located in the south asperity near the initial break point and the asperity in the north of initial break point has a slip of about 2.5 m. The length of the fault plane is approximately 50 km.

DISCUSSION

The Machaze fault plane determined using the MJHD method dips to the west. It is normal fault and the strike, dip and rake (172°, 65° and -78°) of the determined fault plane are adopted from Global CMT. The result based on relocated aftershock distribution is consistent with the field observations by Fenton and Bommer (2006). Those authors observed that the Machaze fault strikes N10-20°W, and the linear trace of the fault indicates that the fault plane is relatively steep, which displacement is normal, down-to-the-west on a west-dipping fault plane. In the rupture process determination, there are two asperities observed. One of them is located at the surface, 15 to 25 km on the north of the initial break point. Source time function (Figure 8c) shows two peaks corresponding to two observed asperities. The second peak occurs five seconds after the first peak which means that the second major slip was observed at the surface five seconds after the major slip. The slip inversion showed also that other than the dominant dip slip offset, there are a small left-lateral offset (Figure 8). This result is consistent with the observations presented by Fenton and Bommer (2006). Aftershock distribution as observed in Figure 9 points that events in the northern side of the mainshock are shallower and in the southern side are deeper. That aftershocks distribution is consistent with the observed asperities from slip inversion. From slip inversion it is observed that the asperity located in the north of the initial break point is shallower and the southern one is deeper. In Figure 9 the fault plane is projected assuming the result of fault plane model determined from the slip inversion. This Figure shows a steeper fault plane (dipping 74°) and it fits well the aftershock distribution reducing the scatter observed when the orientation of fault plane determined by Global CMT is adopted.

Hashimoto *at el.,* (2007) observed from the analyses of ENVISAT images that the source fault is elongated in both north and south directions from the surface rupture, but they couldn't observe clearly the extent in the north of rupture area. In the fault slip distribution presented in this study, it's clear that there is no extension of slip distribution in the north of surface rupture area, related to coseismic rupture.

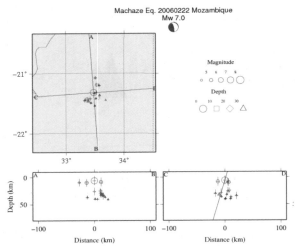

Figure 9. Aftershock distribution of Machaze event adopting the mechanism determined from slip inversion. The symbols are as in Figure 4.

The surface offset determined from the slip inversion in the present study is about 2.5 meters and that obtained by Hashimoto *at el.,* (2007) is about 2.0 ~ 2.7 meters. Those results are consistent each other. The fault length interpreted by Hashimoto *et al.,* (2007) is about 50 km.

The result from aftershock distribution suggests a fault length of approximately 50 km which is the same obtained from slip inversion. Furthermore, all analyses indicate bilateral rupture of the earthquake. The fault plane of the biggest event of Lacerda dip to the east and the strike is approximately N-S, which is consistent with the general strike of the Lacerda graben. Hypocenter distribution of this swarm shows that all of events lie along the same line suggesting that they belong to the same seismic plane.

CONCLUSIONS

Using the MJHD technique, a foreshock, the mainshock and aftershocks of the 2006 Machaze earthquake are accurately relocated in the present study. The fault plane dips to the west with a steep angle. The length of the aftershocks area is approximately 50 km. This study revealed also that the Lacerda cluster is an earthquake swarm. Although it should be an earthquake swarm, all events lie along the same line suggesting that they belong to the same seismic plane which dips to the east. Analyzing the orientation of both Machaze fault plane and Lacerda fault plane determined in this study and relating them with the tectonics of the area where they are located, it can be concluded that they are related to the progressive continental break-up, and intra-continental tectonism related to EARS. From the slip inversion analyze of the Machaze fault, it can be concluded that the Machaze earthquake had two main asperities, one of them located on the surface, which originated an offset of 2.5 meters. The initial break point was not the point of major slip of the fault. The fault slip confirmed that the fault was mainly normal faulting, but there is a small left-lateral slip along the fault plane. The length of the rupture zone is about 50 km and is consistent with the length of aftershock area.

ACKNOWLEDGEMENTS

Expression of gratitude is addressed to Professor Yagi, from University of Tsukuba for the program, and to Prof. Manabu Hashimoto and Prof. Julian Bommer for the valuable manuscripts provided.

REFERENCES

Fenton, C., Bommer, J. 2006. *Seism. Res. Let.* Vol. 77, No. 4, pp 424-439.
Hashimoto, M., Fukushima, Y., Ozawa, T. 2007. *FRINGE*
Hurukawa, N., Imoto, M. 1990. *Journal of seism. Japan, ser 2, 43,* 413-429,
Hurukawa, N, and Imoto, M. 1992, *Geophys. J. Int., 109,639-652.*
Hurukawa, N. 1995. *Geophy. Res. Letters,* Vol. 22, No. 23, pp 3159-3162.
Hurukawa, N. 1998. *Bull. Of the Seism. Soc. America,* vol. 88, No. 5, pp 1112-1126.
Hurukawa, N., Popa, M., Radulian, M. 2008. *Earth Planets Space,* 60, 565-572,
Yagi, Y and Fukahata, Y. 2008. *Geophy. J. Int.,* (In press)

HYPOCENTER RELOCATION AND MOMENT TENSOR ANALYSIS OF EARTHQUAKES IN MYANMAR: TOWARD THE INVESTIGATION OF THE BURMA SUBDUCTION-SAGAING FAULT SYSTEM

Pa Pa Tun[*]
MEE07164

Supervisor: **Bunichiro SHIBAZAKI**[**]
Nobuo HURUKAWA[**]

ABSTRACT

For understanding tectonics and precise earthquake hazard assessment, the detail analysis of hypocenters and moment tensor are necessary. First, present study determined the moment tensor solutions around Myanmar using teleseismic body wave form by the inversion program_by Yagi. Moment tensor solutions obtained in this study are almost consistent with the global CMT solutions. The general orientations of P-axis for shallow earthquakes are consistent with the NNE direction of the Indian plate motion with respect to the Eurasian plate. Next, we relocated hypocenters in and around Myanmar, using the modified joint hypocenter determination (MJHD) method developed by Hurukawa and Imoto. This technique allows us to relocate many events simultaneously in order to improve hypocenter location compared with ISC hypocenters; the MJHD method gave the surprisingly clear features of seismicity. Along the Burma arc, the hypocenters are concentrated on narrow inclined zone. The results show the configuration of the subduction zone. Also we clarified the features of seismicity along the inland fault zone. The Sagaing fault system is located at the area where the cut-off depth of seismicity becomes shallower and the strength of the crust is weaker. Furthermore, we found that the earthquakes are concentrated on the active spreading axis and transform fault in the southern opening region. The results give us new constraints on understanding tectonics in and around Myanmar.

Keywords: Moment tensor inversion, Hypocenter relocation, Burma subduction, Sagaing fault

INTRODUCTION

Myanmar is located at the very active tectonic area which includes the Burma oblique subduction, the Sagaing strike slip fault system and the southern opening region. For understanding the tectonic features of Myanmar, we determined the moment tensors using the body wave form inversion. The focal mechanism bears information on tectonics around the earthquake source region: plate motion and the tectonic stress which causes the earthquake. Seismic moment M_o represents the size of an earthquake. Hypocenters will be useful for determining the subduction interface. Along the Burma subduction zone, there is a high potential for occurrence of large tsunamigenic earthquakes. Therefore, we need to urgently determine the location of the subduction interfaces. The purpose of this study is to confirm the plate boundary between the Indian plate and Eurasian plate along the subduction zone and to understand the tectonic features in Myanmar.

[*]Department of Meteorology and Hydrology (DMH) of Myanmar.
[**]International Institute of Seismology and Earthquake Engineering, Building Research Institute, Japan.

DATA

Moment tensor inversion

We retrieved data from the Data Management Center of the Incorporated Research Institution for Seismology (IRIS-DMC) via internet. In this study, we analyzed only teleseismic body wave (P-waves) from 54 events in the range of Latitude between 8.00° N and 30.00°N and Longitude between 90.00°E and 102.00°E with the body wave magnitude (M_b) above 5.5.

Hypocenter determination

In this study we used the earthquake phase data from International Seismological Center (ISC) catalog from January of 1980 to December of 2004. The data include the P-arrival times from worldwide stations.

THEORY AND METHODOLOGY

Moment tensor inversion

We represented the seismic source process as point source model. We can write the vertical component of the seismic waveform at the station j as

$$u_j(t) = \sum_{q=1}^{6} M_q \int G_{jq}(t - \tau, x_c, y_c, z_c) T(t) d\tau + e_o + e_m \dots\dots\dots\dots (1)$$

where G_{jq} is the complete Green's function, M_q is the moment tensor at centroid of source (x_c, y_c, z_c), $T(t)$ is source time function, e_0 is observation error, and e_m is modeling error. Also we can write this equation in simple vector form as.

$$\mathbf{d} = \mathbf{G}(T(t), x_c, y_c, z_c)\mathbf{m} + \mathbf{e} \dots\dots\dots\dots\dots\dots\dots (2)$$

where $\mathbf{d}$ is the observation waveform, $\mathbf{G}$ is the convolution Green's function with source time function and $\mathbf{m}$ is the moment tensor solution.

Moment tensor inversion needs an assumption of location of hypocenter and information of source time function (duration, shape). We determined the depth of hypocenter and duration and shape of source time function by grid search method. In this study, we used the prem-modify-model velocity structure to find the moment tensor solution. We calculated the Green's function and performed body waveform inversion using programs developed by Yagi (2004).

Hypocenter determination

The equation which is used in the determination of hypocenters is as follows.

$$(O - C)_{ij} = (t_{ij} - To_{oj}) - T_{ij} = \frac{\partial t_{ij}}{\partial x_j}dx_j + \frac{\partial t_{ij}}{\partial y_j}dy_j + \frac{\partial t_{ij}}{\partial z_j}dz_j + dTo_j + dS_i \dots\dots\dots\dots (3)$$

where t_{ij} and T_{ij} are the arrival times and the calculated travel time of the j-th event at the i-th station, respectively. dS_i is a correction of a station correction at the i-th station. To is the origin time. O is the observed travel time. C is the calculated travel time. dx, dy, dz ant dTo are the corrections to the trial hypocenter

Due to the heterogeneous earth's structure, when the station coverage is not good, the JHD solutions become unstable and unreliable because of the trade–off between station corrections and focal depths of earthquakes. For this reason, Hurukawa and Imoto (1990, 1992) developed a modified joint hypocenter determination (MJHD) method using the constraints below.

$$\sum_{i=1}^{n} S_i D_i = 0 \quad \sum_{i=1}^{n} S_i \cos \theta_i = 0 \quad \sum_{i=1}^{n} S_i \sin \theta_i = 0 \quad \sum_{i=1}^{n} S_i = 0 \quad \dots\dots\dots\dots\dots\dots \text{(4)}$$

where S_i is the station correction at the i-th station, D_i is the distance between the i-th station and the center of the region, θ_i is the azimuth to the i-th station from the center of the region, and n is the number of stations.

RESULTS

Moment tensor inversion

In this study, we determined moment tensors to understand various behavior of tectonic and related effect in the region near the subduction and active fault area. We analyzed the 54 events for moment tensor inversion and compared our results (Figure 1) with those from global CMT. Also we used the grid search method developed by Yagi (2004) to find out the best depth and source duration with the minimum variance. Generally almost all the results of moment tensor inversion are consistent with the global CMT results. But some of the results are different in depth and seismic moment from those of global CMT. One of the reasons is that we used the body wave although global CMT is determined using long-period wave. We found that the focal mechanism solution from moment tensor inversion is in agreement not only with global CMT solution but also with the previous studies for regional tectonic setting. According to the generation orientation of P-axis from focal mechanism solution, the direction of P-axis is the same as the Indian plate motion (Figure 2). So this result is consistent with the other results of the Indian plate motion.

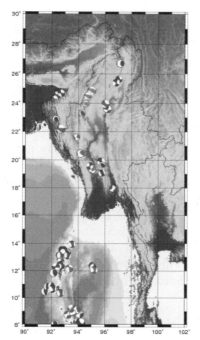

Figure 1. The results of moment tensor inversion.

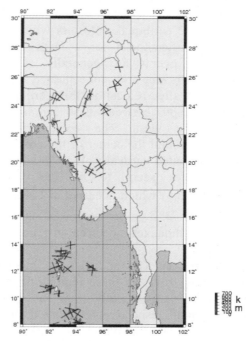

Figure 2. General orientation of P-axis and T-axis.

Hypocenter determination

We relocated the hypocenters for all the events that occurred in and around Myanmar and were obtained from ISC data catalog from 1980 to 2004 including all of the depth range. In this method, to make accurate location, it is better to take a small region with no significant variations in structures. If there is any strong variation in the velocity structures, the station correction strongly depends on the location of events. Therefore, we selected the several small regions for the analysis (Figure 3). Mainly, there are three regions: the Burma subduction zone, the region from the Burma subduction zone to Sagaing fault zone, and the Southern opening region.

Based on the two criteria, we selected the stations and events for each region. These two criteria are minimum number of stations (MSTN) and minimum number of events (MEVN). We remove phase data of which travel time residuals are larger than 2 seconds. To obtain the precise hypocenters, the epicentral distances of nearby stations were checked. There are valuable nearby stations for all events. Therefore absolute locations are reliable for all regions. The station corrections for all of events are almost same.

Seven regions from (A) to (G) are selected for the Burma subduction zone. The region between Latitude from 15.00°N to 24.00°N and Longitude 90.00°E to 96.00°E was included. Figure 4 shows the result in region (A). Before relocation all of events are scattering. (Figure 4a). But after relocation using the MJHD method some of the events are concentrated along the subduction zone and some are related with in inland earthquakes. In Figure 4b, the dotted line are represented the possible plate boundary which are in between Longitude around 91.0°E for this region. However, MJHD method gave the significant of the tectonic features for all regions.

Region (H), (I) and (J) are included in the area from the Burma subduction zone to the Sagaing fault. The tectonic features for those regions are similar each other. Figure 5 illustrates the result in region (I). In this region, we found that the earthquake clusters which seem to be foreshocks and aftershocks for a large event. Before relocation these events are generally at a depth of approximately 10 km and concentrated in horizontal distribution. But after relocation this earthquake cluster appeared at a depth between 0-50 km and down dip direction to the west. The epicenters are much concentrated in around the center of the cross section and seem to be distributed vertically. We can assume that this earthquake cluster seem to be the Taungdwingyi earthquake (M_w 6.6) which occurred at Latitude 19.90°N and Longitude 95.73°E on September 21st 2003. Moreover, four shallow earthquakes occurred just on the Sagaing fault.

The regions (K) and (L) are included in the Southern opening zone. In these regions the earthquakes clusters occurred systematically along the spreading axis and transform fault respectively. Figure 6 show that the MJHD hypocenter distribution for region (L).

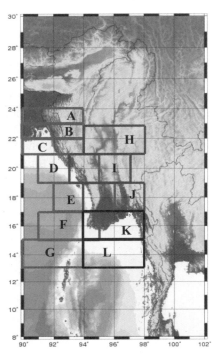

Figure 3. Map showing the 12 sub regions. The rectangulars (A-G) with red color include the Burma subduction zone, the rectangulars (H-J) with blue color include the Burma subduction zone to Sagaing fault and the rectangulars (K-L) with black color include the Southern opening zone.

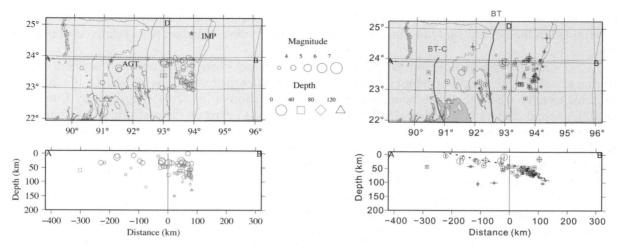

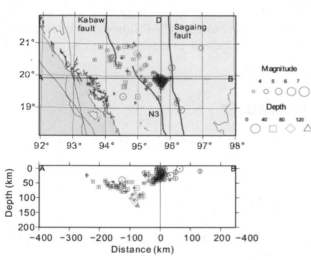

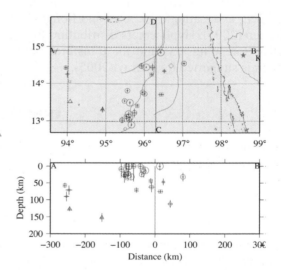

Figure 4a. ISC hypocenter in region (A). The epicenter distribution and two vertical cross sections along A-B and C-D lines are illustrated. A dotted rectangular in epicentral map indicates the epicentral area of selected earthquakes. The size of the symbols represents the magnitude of the events. The color and shape of the symbols denotes the depth range of the events

Figure 4b. Relocated hypocenters by the MJHD method in region (A). Cross represent the standard errors of hypocenters. The symbols are the same as in Figure 4a.

Figure 5. The relocated hypocenters by the MJHD method for region (I). The symbols are the same as in Figure 4b.

Figure 6. MJHD hypocenters for region (L). The symbols are the same as in Figure 4b.

5. DISCUSSION AND CONCLUSION

To investigate the tectonic setting around Myanmar, we obtained moment tensor solutions in and around Myanmar using the teleseismic waveform inversion code developed by Yagi (2004). We analyzed 54 events and compared the results with the solutions from global CMT. Some differences are seen in centroid depth and seismic moment between our result and global CMT solution. Those differences may be due to the difference in method for determining moment tensors. The P-axis determined in the present study is consistent with the NNE motion of Indian plate relative to Eurasian plate.

Since the subduction zone at the shallower part is not clear at the northern segment of the Burma trench. Depth ranges of seimicity become narrower after the relocations by MJHD. It's difficult to judge exactly where the plate boundary is. Around the deformation front suggested by Cummins (2007), seismic activity is quite low. Deformation front seems to exist on the east side of the suggested deformation front. The previous study by Khan (2005) analyzed the hypocenter distribution along the subduction zone using the ISC data between 1964 and 1999. Since the hypocenter distribution was very scattered, they could not get exact images of seismicity along the subduction zone. Our study got much improved hypocenter distribution which is useful for understanding the tectonics in Myanmar.

In the Burma subduction zone to the Sagaing fault, the hypocenters are relocated very well. There are two groups of earthquake hypocenter distributions. One is representing the subduction zone earthquakes, and the other one is related to the Kabaw and Sagaing faults. According to depth distribution of earthquakes, subduction zone earthquakes are located on intermediate depth. On the other hand, earthquakes near the Kabaw and Sagaing faults are located on the shallow depth. Before relocation, we can not clarify the accurate depth of earthquakes clusters. But after relocation, all of these earthquakes clusters are concentrated very well. These events seem to be aftershocks of the Taungdwingyi earthquake which occurred at Latitude 19.90°N and Longitude 95.73°E on September 21st 2003.

As to the Southern opening zone, before relocation, hypocenters of ISC are rather scattering. After relocations, hypocenters are clustering. To clarify the cause of seismic activity around this zone, we compared seismicity with the fault lines which are related with the Sagaing fault (Nielsen, 2004; Curry, 2005). We found that the two seismic activities occurs around Latitude 16.00°N and Longitude 95.70°E, and Latitude 15.00°N and Longitude 95.70°E and these are close to the fault lines drawn by Nielsen (2004). Hypocenters distribution from Latitude 13.00°N to 15.00°N follows systematically the active spreading axis and the transform fault which is the southern extension of the Sagaing fault, which is supported by Curray (2005).

AKNOWLEDGEMENT

I am grateful thank to Professor Yagi (Tsukuba University) for his valuable guidance, suggestion and supporting the programs.

REFERENCES

Bhattacharya, P.M., Pujol, J., Majumdar, R.K. and Kayal, J. R., 2005, Current Science, 89, 1404-1413.
Cummins, P. R., 2007, Nature, 449, doi: 10.1038/nature06088.
Curray, J. R., 2005, Journal of Asian Earth Sciences 25, 187-232.
Hurukawa, N. and Imoto, M., 1990, J. Seism. Soc. Japan, Ser. 2, 43, 413-429.
Hurukawa, N. and Imoto, M., 1992, Geophys. J. Int., 109, 639-652.
Khan, P. K, 2005, Geo. Scie. Journal, 9, 227-234.
Nielsen, C., Chamot-Rooke, N., Rangin, C and the Andaman Cruise Team, 2004, Geology, 209, 303-327.
Rao, N.P., and Kalpna, 2005, Geophys. Res. Lett. , 32, L05301, doi: 10.1029/2004GL022034.
Sahu, V. K., Gahalaut, V. K, Rajput, S., Chadha, R. K., Laishram, S. S. and Kumar, A., 2006, Current Science, 90, 1688-1692.
Socquet, A., Vigny, C., Chamot-Rooke, N., Simons, W., Rangin, C. and Ambrosius, B., 2006, J. Geophys. Res., 111, B05406, doi: 10.1029/2005JB003877.
Yagi, Y., 2004, Earth Planets Space, 56, 311-316.
Yagi, Y., 2006, IISEE Lecture Note 2007-2008, IISEE, BRI.

Synopsis of Master Papers   *Bulletin of IISEE, 43, 31-36, 2009*

ESTIMATION OF SHEAR WAVE VELOCITY STRUCTURE USING ARRAY OBSERVATION OF SHORT PERIOD MICROTREMOR IN KOSHIGAYA CITY, JAPAN

Dayra Yessenia BLANDON SANDINO[*]   **Supervisor: Toshiaki Yokoi**[**]
MEE07161   **Koichi Hayashi**[***]

ABSTRACT

In this study, Spatial Autocorrelation Method (SPAC Method) was carried out in Koshigaya City, Saitama Prefecture, Japan. Additionally, we applied H/V spectral ratio or Nakamura method to the same area.

We obtained that in the western part of the study area the engineering bedrock (Vs>400m/sec) is as shallow as 10-15 m, where as in the eastern part it is much deeper and reaches to 50-60m. The former is categorized in Class D or Stiff Soil of NEHRP (2001) Ground Classification, and the latter in Class E or Soft Soil. We could obtain the same soil classification by implementing the method proposed by Kon'no et al. (2007), which allow us to skip the inversion of the dispersion curve.

Finally, H/V spectral ratio could not show any correlation of its predominant period with the underground velocity structure determined by SPAC method. Furthermore, with the provided information from Nakamura method it is difficult to obtain the ground classification at a site.

Keywords: Microtremor, SPAC method, H/V spectral ratio, S-wave structure.

1 INTRODUCTION

The interaction of the Caribbean and Cocos plates in the subduction area of Nicaragua provokes a great number of earthquakes per year. Many important cities are located in this zone, where mitigation of Seismic Risk is essential. This is the case of Managua, the capital of Nicaragua that has been impacted for two destructive earthquakes in current times (1931 and 1972). This situation led to the elaboration of the seismic microzonation map, by using H/V spectral ratio (Nakamura method). This method, however, is under criticism for its validity and it is suggested that a better result can be obtained by using new techniques of microtremor observation.

Thus, the purpose of this study is to understand the effectiveness, limitations and advantages of the SPAC method using microtremor array measurement as a tool for seismic microzonation and earthquake disaster mitigation by applying them to the study area in Koshigaya City where a complicated underground structure is expected. Nakamura method is conducted additionally.

---

[*] Geoscientific Research Center, National Autonomous University of Nicaragua (CIGEO/UNAN-Managua).
[**] International Institute of Seismology and Earthquake Engineering (IISEE), BRI, Japan.
[***] OYO Technology and Engineering Center, OYO Corporation.

2 DATA

The microtremor field measurements were performed in the urban area of Koshigaya city, approximately in the coordinates: lat. 35° 52' 12''N, lon. 139° 48' 36''E, at the southeast of Saitama Prefecture, approximately 25km from central Tokyo (Figure 1).

The measurements were conducted at 12 sites, on June 12 and 13, 2008 in day time. The arrays were deployed in the shape of the letter "L" for SPAC method, which consists of eleven geophones (Natural frequency 2Hz) and simultaneously, we used the three-component seismometer GEO-SPACE LP of Geometrics (Natural frequency 1Hz) for H/V spectral ratio at each site.

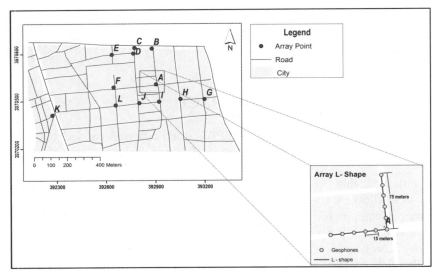

Figure 1 Map showing array configuration adopted in microtremor measurements. The test field is located in Koshigaya City, Saitama Prefecture, Japan. The box in the right-down side indicates the L-shape array with the distance among sensors.

3 METHOD

3.1 SPAC Method

The SPAC Method is based on the theory developed by Aki (1957) to comprehend the relationship between the temporal and spatial correlation of seismic waves and microtremors, which it became the key to successful extraction of dispersion characteristics of Rayleigh waves. (Okada, 2003).

The SPAC coefficients denoted by equation (1) can be directly calculated in a frequency domain using the Fourier Transform of the observed microseism, that is,

$$\rho(\omega;r) = \frac{1}{2\pi} \int_0^{2\pi} \frac{real[S_{CX}(\omega;r,\theta)]}{\sqrt{S_{cc}(\omega;0,0) \cdot S_{xx}(\omega;r,\theta)}} \tag{1}$$

where real $[\cdot]$ stands for the real part of a complex value, and $S_{cc}(\omega;0,0)$ and $S_{xx}(\omega;r,\theta)$ are the power spectra of the microseism at two sites, $C(0,0)$ and $X(r,\theta)$, respectively. $S_{cx}(\omega;r,\theta)$ is the cross spectrum between $u(t;\omega,0,0)$ and $u(t;\omega,0,\theta)$. The two autocorrelations in the denominator make the local amplification effect canceled out (Okada, 2003).

Then, the velocity that makes the misfit function minimum is searched for each available frequency using all considered inter-station distance. The obtained value of c is considered as the phase velocity at the corresponding frequency. After that, we obtained the optimum underground structure for the given dispersion curve of Rayleigh wave based on the Down Hill Simplex Method (DHSM) combined with the Very Fast Simulated Anealing (VFSA) approach (Yokoi, 2005).

The initial model used has fives surface layers and a half space that represents the bedrock of which Vs is about 400 m/sec (Table 2). For making this initial model, the information obtained from a borehole GS-SK-1A was considered (Ishihara, 2004; Hayashi et al., 2006).

Table 2 Initial Model for all the sites

h_{min} (Km)	h_{max} (km)	Vs_{min} (km/sec)	Vs_{max} (Km/sec)
0	0.03	0.08	0.15
0.001	0.03	0.1	0.15
0.001	0.03	0.08	0.15
0.001	0.03	0.15	0.25
0.001	0.03	0.25	0.35
998	999	0.35	0.45

The Vp and the density ρ are calculated using the following empirical formulas for each step of iteration (Ludwig et al.,1970; Kitzunezaki et al., 1990, respectively).

$$Vp=1.11\ Vs+1.29 \quad (Km/sec) \tag{2}$$

$$\rho=1.2475+0.399\ Vp-0.026\ Vp^2 \quad (g/cm^3) \tag{3}$$

3.2 Nakamura method

Nakamura (1989) proposed a method of inferring site amplification factors from incident seismic shear waves using microtremor H/V spectral ratios at a single site. This method is easily applied and directly estimates the site amplification factors without reference site, and many researchers have done to investigate the validity by observation and in theory. (e. g., Horike et al., 2001; Lermo and Chavez-Garcia, 1994).

The calculation is done by the following formula

$$H/V_{spectralratio} = \sqrt{\frac{\sum P_{NS}(\omega) + \sum P_{EW}(\omega)}{\sum P_V(\omega)}} \tag{4}$$

where $P_{NS}(\omega)$, $P_{EW}(\omega)$ and $P_V(\omega)$ are the power spectra of NS, EW and the vertical component respectively, summation is taken over data blocks.

4 RESULTS AND DISCUSSION

4.1 SPAC method

We obtained the dispersions curves or phase velocity for each observation site. The result indicates phase velocities lower than 0.5 (Km/sec) at all the sites and suggests that the region is covered by soft sediments up to the depth that can be explored in this research (Figure 2).

Figure 2 shows the result of dispersion curve for all the sites except the site G where unusual phenomena occurred in SPAC coefficients. Namely, the usual order, in which those of longer inter station distance, has smaller value in the range from $kr=0$ to the first zero cross, is not observed.

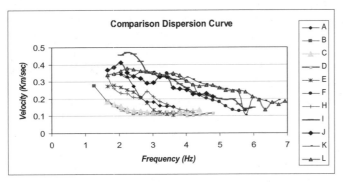

Figure 2 Dispersion curves for all sites except the site G.

This may be interpreted as the failure of the assumption: horizontal stratified layers or unrecognized human made noise. Therefore, the site G is excluded from the analysis.

The sites can be separated into four groups by the similarity of the shape and values of the dispersion curves. Namely, *Group 1* corresponds to the sites B, C and D. *Group 2* corresponds to the sites A, E and H. *Group 3* corresponds to sites F, I, and J. *Group 4* corresponds to the sites K and L.

After obtaining dispersion curve, the result was inverted to Vs structure. Figure 3 shows Vs structure for all the sites except the site G. It is evident that even though we conduct array measurements in a small area, a clear variation exists in Vs structures among the observation sites.

For *Group 1*, we found the engineering bedrock (Vs about 0.400 (Km/sec)) at depth of 0.055 (Km), except at the site D where it appears at about 0.040 (Km); the layer with Vs approximately 0.200 (Km/sec) appears at depth about 0.030 (Km). For *Group 2*, the depth of the engineering bedrock varies from 0.030, 0.024 and 0.053 (Km), at the site A, E, and H, respectively; the layer with Vs approximately 0.200 (Km/sec) appears at depth of about 0.15 (Km/sec), but at the site A appears at 0.022 (Km). For *Group 3*, the engineering bedrock was found at depth from 0.015 to 0.018 (Km); the layer with Vs approximately 0.200 (Km/sec) appears at depth of about 0.008 (Km). Finally *Group 4*, the engineering bedrock was found at depth from 0.010 to 0.015 (Km); the layer with Vs approximately 0.200 (Km/sec) appears at depth less than about 0.008 (Km).

We can judge clearly that *Group 1* located at north-east part of the study area corresponds to the thickest sediment, while *Group 4* located at the south-west part corresponds to the thinnest and that the thickness of the soft sediment is decreasing from north-east to south-west.

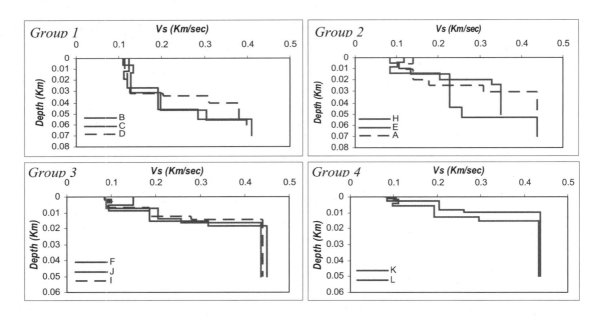

Figure 3 S-wave structures of four groups.

4.1.1 Soil profile

By using the result about Vs structure, we drew two soil profiles. The *profile one* connects the sites KFDB with direction south-west to north-east, and the *profile two*, connects the sites KLJIH from west to east in a straight line. (Figure 4)

The *profile one* shows that, the depth change considerably with difference around 23 m in a distance of 246 m between site F and D. After this point, the depth increases up to about 55 m at the site B. Here, the sub-area at west side of the site F including the site K may correspond to the buried terrace, while at north – east of the site F may correspond to the slope between buried terrace and the buried channel.

For the *profile two*, the depth change considerably between site I and H (136.33 m of distances between these sites) which is significant for the layers four and five. The maximum depth reached is about 53 m for layer 5.

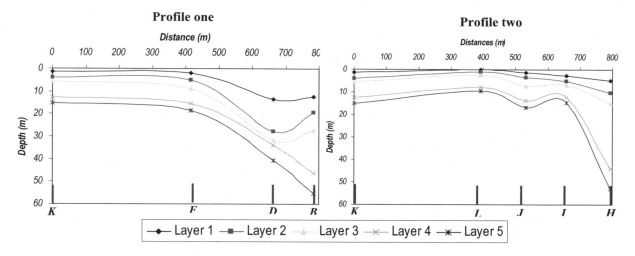

Figure 4 Soil profile along the sites K-F-D-B (left) and K-L-J-I-H (right).

4.1.2 Ground classification

By using the result from SPAC method, the ground classification is conducted based on AVS30, based on the following equation:

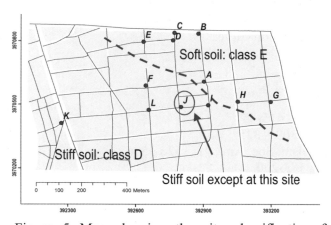

Figure 5 Map showing the site classification for observation sites, indicating soft soil at the northeast and stiff soil at the southwest of the study area (except at the site J that is soft soil).

$$AVS\,30 = \frac{\sum_{i=1}^{n} d_i}{\sum_{i=1}^{n} \frac{d_i}{Vs_i}}, \qquad (5)$$

where Vs_i is the shear wave velocity in m/sec, d_i the thickness of *i-th* layer between 0 and 30 m. (NEHRP, 2001).

The result was Soft soil (Class E: AVS30< 180 m/sec) at the north-east parts (the sites A, B, C, D, E, H, and J) and Stiff soil (Class D: 180 m/s ≤ AVS30≤ 360 m/s) at the south-west (the sites F, I, K and L). Even though it is not visible on the topography, there is a lateral variation of the bedrock depth. As

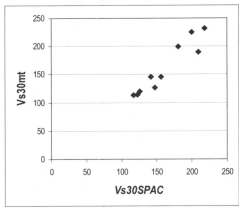

Figure 6 Comparison of Vs30mt with Vs30SPAC

the calculated values for the sites I and J are close to the threshold, it is better to consider that these two sites are in transition between Class D and Class E (Figure 5).

Also, we use the method proposed by Kon'no et al. (2007) as an approximate way to estimate AVS30 without performing the inversion of the dispersion curve. They have proposed the use of the phase velocity provide by SPAC method for the wavelength 40m as an approximation of AVS30 (Vs30mt).

The result was the same as that of the previous method. Only for the case of the site J of which value was not found due to limitation of the available frequency. However, Vs30mt can be guessed less than 213 (m/sec) for this site at least. In Figure 6 the comparison between these two methods is shown, of which deviation is acceptable.

4.2 H/V spectral ratio (Nakamura Method)

The sites in *Group 1* show a clear peak of predominant period less than 1sec. For the rest of Groups, the predominant period is longer than 1 sec, except at the site A where was around 0.9 sec.

Taking into account the soil classification provide by SPAC method, we can say that the longer period corresponds to Stiff soil class, and shorter one is found at Soft soil class. This is opposite of the relation of the predominant period of amplification of up-coming S-wave. The result is an evidence of the failure of Nakamura method in the study area, which predominant period of amplification is supposed to be shorter at Stiff soil and longer at Soft soil.

5 CONCLUSION

SPAC method could give a quantitative estimate of shear wave velocity structure that is consistent with previous information about the depth of bedrock and geology of the area. Namely, in the western part of the study area the engineering bedrock (Vs>400 (m/sec)) is as shallow as 10-15 m, where as in the eastern part it is much deeper and reaches to 50-60 m. The former is categorized in Class D or Stiff Soil of NEHRP (2001) Ground Classification, and the latter in Class E or Soft Soil.

We could obtain the same soil classification by implementing the method proposed by Kon'no et al. (2007), which use phase velocity for the wavelength 40m as an approximation of AVS30 (Vs30mt), without performing the inversion of the dispersion curve.

REFERENCES

Aki, K., 1957, Bull. Earthq. Res. Inst. Univ. Tokyo, 35, 415-457.
Hayashi, K., et al., 2006, Bull. Geol. Surv. Japan, 57, 309 – 325.
Horike, M. , B. Zhao and H. Kawase , 2001, Bull. Seism. Soc. Am., 91, 1526-1536.
Ishihara, Y.et al., 2004, Bull. Geol. Surv. Japan, 55, 183-200.
Kitzunezaki, C., et al., 1990, Jour. of Japan Soc. for Natural Disaster Science, 9-3, 1-17
Kon'no, K., et al., 2007, Jour. of Japan Society of Civil Engineering, 63, 639-654.
Lermo, J and Chavez-Garcia F. J, 1994, Bull. Seism. Soc. Am. 84, 5, 1350-1364.
Ludwig, W. J., et al., 1970, Seismic refraction, The Sea 4-1, 53-84.
Nakamura, Y, 1989, QR of RTRI 30, no 1, Februrary, 25-33.
NEHRP, 2001, Part 1. Provisions (FEMA 368), Washington DC. Building Seismic Safety Council.
Okada, H., 2003, Geophysical Monograph Series No. 12, Society of Exploration Geophyscists, Tulsa.
Yokoi, T., 2005, Programme and abstracts, The Seismological Society of Japan, Fall meeting.

Synopses of Master Papers   *Bulletin of IISEE, 43, 37-42, 2009*

TRAVEL TIME RESIDUALS FROM THE NEW AND OLD SEISMIC NETWORKS OF PAKISTAN AND PRELIMINARY HYPOCENTER DETERMINATION

Nasir Mahmood*   **Supervisor: Tatsuhiko HARA** **
MEE07163

ABSTRACT

The Pakistan Meteorological Department has started establishing a new digital seismic network in 2006. We evaluated data from this new network, and performed preliminary hypocenter determination. First, we checked detectability of new stations to find that the detection capability of the Islamabad station seems better than the Peshawar and Quetta stations in new network. Then, we calculated theoretical P and S wave travel times using USGS hypocenters. We compared them to the observed travel times from the new and old Pakistan seismic networks to obtain travel time residuals. For station Quetta in the old analog network, the P and S readings are less accurate than those from the new network. Taking these uncertainties into account, we performed preliminary hypocenter determination of events in the Hindu Kush region using data with relatively smaller residuals. We used two earth models: the one is iasp91 and the other has a crust structure from CRUST 2.0. Most of the events are located in the depth range consistent with USGS focal depths when we used iasp91, while the distances between the epicenters issued by the USGS and those determined using iasp91 are larger (on the order of 50 km) than those when we used the latter model (on the order of 25 km). Probably due to the small number of data and not good station configuration to events in the Hindu Kush region, the dependence of hypocenters on crust models is strong. We expect it will be possible to improve accuracy of hypocenters by expanding our seismic network with carefully checking data.

Keywords: New digital seismic network, Travel time residuals, Hypocenter determination, Hindu Kush Region.

INTRODUCTION

The seismic networks in Pakistan

An advanced earthquake monitoring system is necessary for countries such as Pakistan surrounded by the active seismic zones. At the time of the October 8th 2005 Kashmir earthquake, there was a small analog seismic network of six stations including one digital station at Peshawar, which were operated by the Pakistan Meteorological Department. The performance of this network was not sufficient, and a denser seismic network with modern communication and recording systems is necessary to enhance the monitoring capability.

The Pakistan Meteorological Department has deployed a set of new broadband Guralp sensors (CMG 40T) at Islamabad, Peshawar, Quetta and Balakot in 2006. *Scream 3.0* software was installed for seismic data acquisitioning on a real time basis at each station. Seismic signals

---

*Pakistan Meteorological Department, Ministry of Defense, Pakistan.

** International Institute of Seismology and Earthquake Engineering, BRI, Tsukuba, Japan.

were digitized at 24 bits with a sampling rate 100 sps at each station. Both Islamabad and Peshawar stations are linked via telephone line to exchange the seismic data, while other stations were under consideration to be linked. This new seismic network is expected to enhance the earthquake monitoring capability (although it is desirable to increase the number of stations particularly in areas close to seismically active zones). In this study, we evaluated data from this new network, and performed preliminary hypocenter determination. First, we checked event detectability of the new stations.

Event detection capability and data quality of new network

The Islamabad station CMG 40T sensor in new network has recorded relatively a larger number of events, which suggests its better detection capability. The smallest event of magnitude M=3.3 detected at epicentral distance 120 km. An event (M=6.1) with epicentral distance of 728km in the north direction was recorded with clear P and S phases. Detection capability of CMG 40T sensor at Peshawar station was poorer than the Islamabad station. Some of the events in the Hindu Kush region recorded at Islamabad but can not be recorded at the Peshawar station, although they are closer to Peshawar than to Islamabad. At Peshawar, another sensor, S-13, has been deployed, and its detection capability is much better. Many events that are not detected by CMG 40T sensor are recorded by this sensor. Due to unstable power condition, instrument handling and some technical problems only a few events were found from the observed records of CMG 40T sensor of Quetta station. The detection capability of this station is very poor compared to the Peshawar and Islamabad stations. All of the events recorded by the Islamabad, Peshawar and Quetta stations between 2006 and 2007 are shown in Fig. 1.

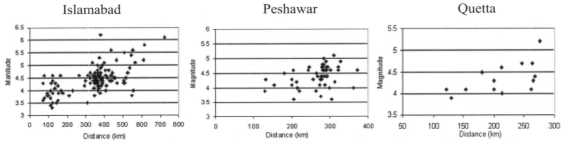

Figure 1. Events detected by CMG 40T sensors at Islamabad, Peshawar and Quetta stations

However the WWSSN system at Quetta recorded more events, and we used these data in the following part of this study. Similarly quality of seismic data from the Islamabad station was found comparatively better than those of the other stations in new network. Noise level of the observed records of Islamabad (CMG 40T) was quite low. A local event of M=3.5 was recorded with clear arrivals of P and S waves. For Peshawar station (CMG 40T), larger noise level observed and it was difficult to pick P arrivals for smaller events of magnitude M ≤ 4.0 only S phase was clear. Similarly it was difficult to pick S phase for smaller events from the CMG 40T records of Quetta station. A possible problem of noise levels may come from the local site condition of the sensor for both these stations.

TRAVEL TIME RESIDUALS FROM NEW AND OLD SEISMIC NETWORKS

In this study we used data from the new network of Guralp sensors (CMG 40T) at Islamabad and Peshawar stations, S-13 sensor of Peshawar station and the old WWSSN system of Quetta station for the period 2006-2007. The main aim was to check the accuracy of data from the observed record of the new and old networks and investigate factors causing discrepancies in the data. We used the SEISAN version 8.1 (Havskov and Ottmoller, 2005) software to manipulate and analyze the digital waveform data. The digital data obtained from the new network was first converted from Guralp (GCF) to SEISAN format. Arrival times for P and S waves were picked for all possible events from each station. Observed travel times for P and S waves were obtained by subtracting origin times (issued by the USGS) from observed arrival times for each station. Theoretical travel times for P and S waves were calculated using TauP toolkit software (Crotwell et al., 1999) and iasp91 model (Kennett and Engdahl, 1991). Travel time residuals for P and S waves as well as S-P times residuals were computed using the observed and theoretical travel times. In this way we compared observed and theoretical travel times to check the accuracy of data from the new and old networks and found following results for three stations Islamabad, Peshawar and Quetta.

Data from Islamabad station

Figure 2(a) shows a comparison between observed and theoretical P wave travel times for the Islamabad station. There was a general agreement between them, while they differ considerably for some events. Figure 2(b) shows the comparison for events whose magnitudes are equal to or greater than 4.5. Most of the data which show large differences are removed. This suggests that there are many misidentifications for smaller events. We made similar comparisons for S travel times and S-P times and find a general agreement between observed and theoretical times.

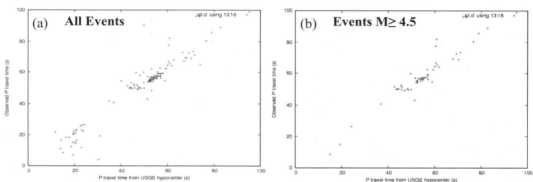

Figure 2. Comparison of observed and theoretical P wave travel times for Islamabad station.

Figure 3 shows the frequency distributions both for P and S travel times and S-P times respectively for the Islamabad station. The P travel time residuals show a peak around 2 sec and most of the observations are within the range of 0 – 5 sec. The S travel time residuals show a larger scatter, and most of the observations are within the range of -3 – 3 sec.

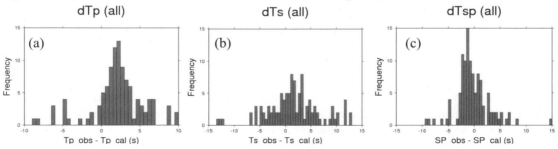

Figure 3. Frequency distributions of P travel time residuals (a), S travel time residuals (b), S-P time residuals (c), respectively, for Islamabad.

Since there is not a large systematic deviation from 0 both for P and S wave travel time residuals, it is possible to use absolute travel times in the following hypocenter determination.

Data from Peshawar station

Figure 4 shows the frequency distributions of the P and S travel time residuals and S-P time residuals for Peshawar station (CMG 40T). P wave travel time residuals show a peak around 2 sec ad most of the observations are within the range of -1 – 5 sec while S wave travel time residuals show a larger scatter and most of the observations are within the range of -5 – 4 sec. S-P time residuals are within the range between -5 – 2 sec. Similarly we used absolute travel times from this station in the following hypocenter determination.

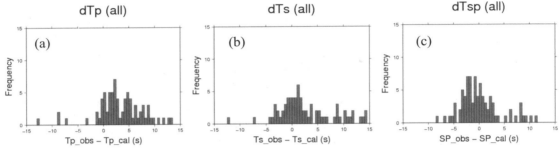

Figure 4. Frequency distributions of P travel time residuals (a), S travel time residuals (b), S-P time residuals (c), respectively, for Peshawar.

Data from Quetta station

Figure 5 show the frequency distributions of the P and S travel time residuals and residuals of the S-P times for WWSSN of Quetta station. Both P and S wave travel time residuals show larger scatters. S-P time residuals are comparatively better. We preferred S-P times from this station.

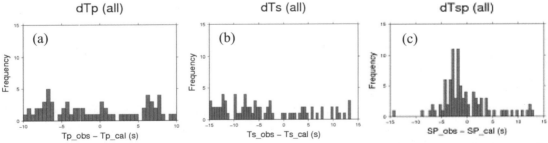

Figure 5. Frequency distributions of P travel time residuals (a), S travel time residuals (b), S-P time residuals (c), respectively, for Quetta.

HYPOCENTER DETERMINATION

In installation of the new digital network, the primary emphasis was to increase the number of stations. Seismic data analysis was mostly carried out manually including earthquake location, and sometimes manual results were not satisfactory or very poor. One of the main objectives of this study was to improve the techniques of hypocenter determination to obtain better results. Since data for an event from all the three stations of the new network were not available, we used data from the analog network as well. Arrival time data from the Islamabad, Quetta and Peshawar stations were used for hypocenter determination.

Software

We used earthquake analysis software SEISAN 8.1 (Havskov and Ottemöller, 2005). After putting data into SEISAN database, we converted waveform data format from Guralp format (GCF) to SEISAN format. Mulplt command is used to plot waveform data for picking P and S phases.

We used HYPOCENTER 3.2 program (Lienert et al., 1986) under SEISAN environment for hypocenter determination. This program has the capabilities to locate earthquakes locally, regionally and globally. Least-squares method works behind this program, in which arrival-time residuals are minimized to get a new set of location parameters. The procedure is repeated through an iterative process till an acceptable error criterion is met. The final adjusted parameters are then accepted as the best possible estimate of the source location.

Crustal Structure

Because there is no velocity model which has been built beneath the country, and most of the analysis carried out manually before, it is necessary to search for a velocity model to carry out earthquake location. We used two models: the one is iasp91 (Kennett and Engdahl, 1991) and the other is the P6 type from CRUST 2.0 (Bassin et al., 2000). This model is available at http://mahi.ucsd.edu/Gabi/rem.dir/crust/crust2.html. Initial guess for hypocenter was set to 15 km.

Analysis and Results

We selected 12 events of the Hindu Kush region for the period 2006-2007 and performed hypocenter determination using arrival time data for P and S waves from the new and old seismic networks of Pakistan. As for data from the Quetta station, based on the results of travel time residuals, we used S-P times. Figure 6 shows comparison between the hypocenters determined using the P6 type from CRUST2.0 and iasp91 with those determined by the USGS.

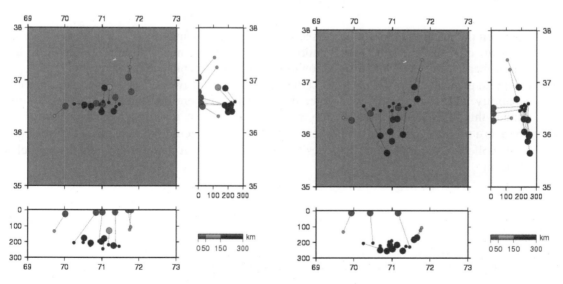

Figure 6. Comparison between the hypocenters obtained for the P6 type from CRUST2.0 (left) and for iasp91 model (right) with those determined by the USGS. Larger and smaller circles denote hypocenters of this study and the USGS, respectively.

The focal depths of the analyzed events determined by the USGS are in the range from 100 to 250 km. Most of the events are located in this range when we used iasp91 model, while the distance between the epicenters issued by the USGS and those determined in this study using iasp91 are larger (on the order of 50 km) than those when we used the P6 mode (on the order of 25 km). Probably due to the small number of data and not good station configuration to events in the Hindu Kush region, the dependence of hypocenters on crust models is strong. We expect it will be possible to improve accuracy of hypocenter determined by expanding our seismic network with carefully checked data.

CONCLUSIONS

In this study, first, we investigated detectability of events. Our results suggest that the detection capability of the Islamabad station seems better than those of the Peshawar and Quetta stations in new network. Then, we calculated theoretical P and S wave travel times using hypocentral parameters of USGS and compared them to those calculated using data from the new and old Pakistan seismic networks to obtain the distributions of travel time residuals. For station Quetta in the old analog network, the P and S readings are less accurate than those from the new network, which resulted in larger travel time residuals.

Considering these uncertainties, we performed preliminary hypocenter determination of events in the Hindu Kush region using data with relatively smaller residuals. When we used iasp91 as an earth model, most of the events are located in the depth range consistent with the USGS estimates. The distance between the epicenters determined by the USGS and those determined using iasp91 are larger (on the order of 50 km) than those when we used the model from CRUST 2.0 (on the order of 25 km). Probably due to the small number of data and not good station configuration to events in the Hindu Kush region, the dependence of hypocenters on crust models is strong.

RECOMMENDATIONS

Based on the results of this study, we recommend the followings.

- The sites of Peshawar and Quetta stations should be examined with appropriate noise level and further site survey while installation of new stations to enhance the quality of data.
- Continuous evaluation of data from stations of the new network will enhance the detection capability and accuracy of hypocenter determination. Therefore, such evaluations as were done in this study should be applied to stations which will be installed in the near future.
- There is a timing problem at the Peshawar station, which should be checked and corrected.
- After collecting enough data, it is desirable to construct an appropriate velocity model for Pakistan

REFERENCES

Bassin, C., G. Laske, and G. Masters, 2000, EOS Trans AGU, 81, F897.
Crotwell, H.P., T.J. Owens, and J. Ritsema, 1999, Seismol. Res. Lett., 70, 154-160.
Havskov, J., and L. Ottemöller, 2005, Univ. Bergen, Norway.
Kennett, B.L.N. and E.R. Engdahl, 1991, Geophys. J. Int., 105, 429-465.
Lienert, B.R., E. Berg, and L.N. Frazer, 1986, Bull. Seis. Soc. Am., 76, 771-783.

Synopses of Master Papers

*Bulletin of IISEE, **43**, 43-48, 2009*

DETERMINATION OF EARTHQUAKE PARAMETERS USING SINGLE STATION BROADBAND DATA IN SRI LANKA

S.W.M. SENEVIRATNE*
MEE07166

Supervisors: Yasuhiro YOSHIDA**
Tatsuhiko HARA***

ABSTRACT

We determined epicenters and magnitudes using data from the single broadband seismic station in Sri Lanka, PALK. First, we analyzed earthquakes to which epicentral distances from PALK were within 40 degree. Most of the events are in and around Andaman and Nicobar Islands in the Indian Ocean. The obtained results show that it is possible to determine epicenters and surface wave magnitudes with acceptable accuracy using a single station method.

We also analyzed an earth tremor in Sri Lanka using data from the PALK station. Despite high frequency noises contained in the data, the epicenter determined by the single station method is consistent with felt reports. The calculated local magnitude is 3.2.

Keywords: Single station method, Epicenter, Surface wave magnitude.

INTRODUCTION

In Sri Lanka, we have not faced with many natural disasters except for landslides based on the known history. In a several decades, there have been implemented preparedness and preventive measures for landslides considering that they are possible natural disasters in the country. When it comes to earthquakes, it is said that Sri Lanka is considered fairly safe and not likely to be prone to major earthquakes. This idea has been changing recently in the society after the Great Tsunami in 2004. Although we are considerably far away from an active plate margin of Sumatra, we suffered from the historical tragedy a several hours after the 2004 M=9 earthquake occurred. This incident proved that seismic activity in subduction zones can transmit a disastrous tsunami for more than two thousand kilometers away from its origin in the Indian Ocean. The seismicity of this area is likely to affect the internal peacefulness of earth in the country. Many off shore earthquakes have made significant shakings inside the country according to the reports after 2004 great earthquake.

A number of inland tremors have been reported by general public within the country in past few years. Most of these tremors were very local, confined to particular region within the country (Fernando et al. 1986). Monitoring of such micro seismic activities is very important and, since there have been recorded a number of tremors within the country, it is essential for us to learn and study local earth dynamics and seismicity.

The great Indian Ocean Tsunami in 2004, reminds us that, although Sri Lanka is located in a relatively stable area, the secondary events like tsunamis can make significant damages. Therefore it is expected to study the seismicity of that particular region using available data resources.

* Geological Survey and Mines Bureau, Sri Lanka.
** Meteorological Research Institute, JMA, Tsukuba, Japan.
*** International Institute of Seismology and Earthquake Engineering, BRI, Tsukuba, Japan.

PALK is the single reliable seismic station in Sri Lanka, data from which was used in the study. We used the software DIMAS to analyze waveform data from 39 regional and teleseismic events and one local event.

DATA

We selected 39 earthquakes recorded at PALK whose moment magnitudes are larger than 6. Although we are mainly interested in events in and around Andaman and Nicobar Islands, some events in other regions are also included in our dataset.

The basic information about events were initially collected using JWEED developed by IRIS data center. The selected events can be further reviewed for the purpose of the study. Once we decide the events of the study, the electronic request is generated and sent to the data center through JWEED. A few minutes later, the requested waveform data are obtained on the client computer and we can save them as mini seed or SAC data formats according to our propose.

METHOD

We used S-P time differences and an assumed velocity structure of the region to find epicentral distances. The P waves and S waves were picked from the vertical component and two horizontal components, respectively. Appling these results with appropriate regional P and S wave velocities in Wadatii diagrams, we can find the epicentral distances quickly. The particle motion of a P wave recorded in all three components can be used to determine the direction of the P wave. Surface wave magnitudes (Ms) were measured using Raleigh waves that were recorded on vertical components. Furthermore, owing to the availability of waveform data, Local Magnitude (M_L) was calculated for a recent local earth tremor.

The software DIMAS (Display, Interactive, Manipulate and Analysis of Seismogram) is used in determination of earthquake parameters in this study (Droznin, 1997). The program has been developed based on the following algorithms.

Origin time

The equation (1) shows the relationship of origin time with arrival time of P wave, S-P time difference and ratio of the velocity.

$$To = Tp - Ts\text{-}p/V, \tag{1}$$

where

To = Origin Time,
Tp = Arrival Time of P wave,
Ts-p = Difference of arrival times of S-wave and P-wave,
V = (Vp /Vs) -1,
Vp = Velocity of P wave,
Vs = Velocity of S wave.

Epicentral Distance

After finding the P wave travel time, assuming the source depth, the epicentral distance is determined by the travel time table. The program uses the default value for depth, which is set to zero. Depending on waveforms, a user can assume the depth based on depth phases as well. The program uses the travel time table for model iasp91 (Kennett and Engdahl, 1991), and the epicentral distance is calculated based on this table.

Source Azimuth

The azimuth from the station to the seismic source is determined by the investigation of polarization of the initial motion of the P waves. Analyzing the particle motion of the P wave in the first few seconds of the three component records and assuming the lateral homogeneity of the region, the direction of the P wave can be obtained.

Magnitude Calculation

Surface and local magnitudes are calculated by the following formula.

Surface wave Magnitude $\quad \mathbf{M_s} = \mathbf{log_{10}(A/T)} + \mathbf{1.66\{log_{10}(delta)\}} + \mathbf{3.33} \qquad (2)$

    A – The maximum amplitude of the surface wave (μm)
    T – The dominant period of the measured wave (seconds)
    delta – The distance from the station to epicenter (degree)

Local Magnitude $\quad \mathbf{M_L} = \mathbf{log_{10}(A/T)} + \mathbf{2.56\{log_{10}(dist)\}} + \mathbf{0.67} \qquad (3)$

    A – The maximum amplitude (μm)
    T – The dominant period of the measured wave (seconds)
    dist – The distance from the station to source (km)

RESULTS

Epicenters

Figure 1 shows the epicenters determined in this study and those from the PDE catalog. We found that the quality of PALK data were good enough for epicenter determination. We calculate distances between the corresponding epicenters of this study and the PDE catalog, and their frequency distributions are shown in Fig. 2. The distances for most events are in the range between -1 and 1 degree, from which we may expect relatively good estimates for magnitudes.

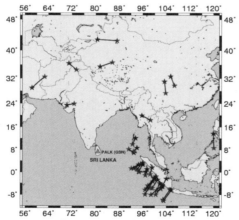

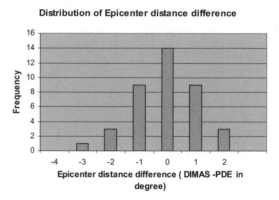

Figure 1. The epicenters determined in the study (red stars) and those from the PDE (blue stars) catalogues are shown, respectively. Corresponding epicenters are connected by solid lines.

Figure 2. Frequency distribution of distances between epicenters determined in this study and those by the USGS.

Magnitude calculation

Figure 3 shows the comparison between Ms determined in this study and those from the PDE catalog, and Fig. 4 shows the distribution of their differences. We find a good agreement between them for the magnitude range between 6 and 8. Above M=8, their agreement becomes poorer. Their magnitude differences are within ±0.2 for most of the events.

Figure 5 shows the comparison between Ms determined in this study and Mw from the Global CMT catalog, and Fig. 6 shows the distribution of their differences. Again, we find a good agreement between them for the magnitude range between 6 and 8. Their magnitude differences are within ±0.3 for most of the events.

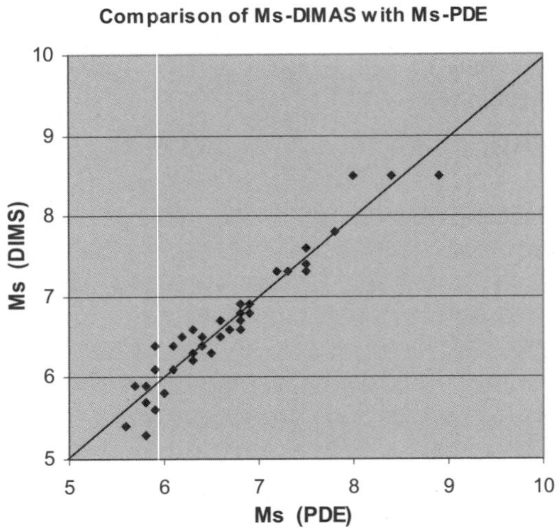

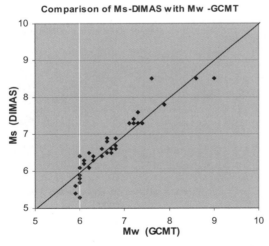

Figure 3. The comparison of Ms obtained from the Study and those from the PDE catalogue.

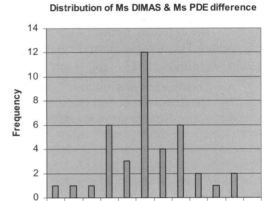

Figure 4. Frequency distribution of differences Ms in the study and those from the PDE catalogue.

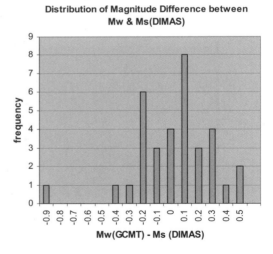

Figure 5. The comparison of Ms obtained in this study and Mw from the Global CMT catalogue.

Figure 6. Frequency distribution of differences between Ms of this study and Mw from the Global CMT catalogue.

Analysis of a local tremor

On the April 7, 2008, there was an earth tremor in the central part of Sri Lanka. According to reports from local authorities and public, shakings were felt around Dambulla and Sigiriya in Central Province.

We applied high band-pass filter to the waveform data from PALK, and found a signal around the time where the shakings were felt. This is likely to a record of the earth tremor. Around the time when this local event occurred, there was another regional event in the Indian Ocean. Therefore, the data of the local tremor data has noises of the considerable level. However, using high band-pass filtering, we could analyze this local event.

The latitude and longitude of the estimated epicenter was 7.84 degree and 80.75 degree, respectively. This result is consistent with the felt reports issued by local authorities. The calculated local magnitude of this event was 3.2.

Back azimuths

Figure 7 shows the distribution of differences between the back azimuths in this study and those calculated using the epicenters from the PDE catalogue. They distribute around 10 degree. One of the possibilities for this discrepancy is laterally heterogeneous earth structure. However, this is not likely in this case, because the differences of the back azimuths are observed not only for the events around Andaman and Nicobar Islands but also for the events in other regions.

Another possible reason is inaccurate information on configuration of the seismometer. It will be useful and interesting to use an artificial event, the precise epicenter of which we know, to determine the actual directions of the sensors.

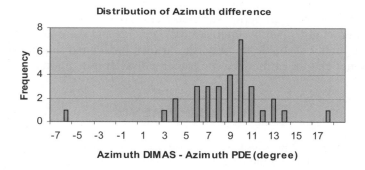

Figure 7. The differences of azimuth values of catalogue (PDE) and calculated in DIMAS

CONCLUTIONS

In this study, we investigated uncertainty of epicenters and surface wave magnitudes determined by a single station method using broadband data from the PALK station. The epicenters obtained in this study agreed well with those of the PDE catalogue in general. The distances between their corresponding events are between −1 and 1 degree for most of the events. We observed the systematic difference of the back azimuths. The plausible reason is likely to be inaccurate information on configuration of the seismometer.

We compared Ms determined in this study to those from the PDE catalogue and Mw from the Global CMT catalogue. In both cases, they agreed well in general in the magnitude range between 6 and 8. Above the magnitude of 8, their differences become larger. One of the reasons may be saturation of Ms for huge events. However, this initial information will be useful and beneficial for tsunami disaster mitigation, because the estimated Ms for the events whose moment magnitudes are larger than 8 are larger than 8, based on which we can consider the possibility of occurrence of devastating events.

We analyzed one local event on the April 7, 2008 using the single station method. Although the data contained considerable noises, the determined epicenter was consistent with felt reports. The calculated local magnitude was 3.2.

RECOMMENDATIONS

The study showed that it was possible to determine epicenters and magnitudes by the single station method using data from PALK with relatively good accuracy. However, uncertainty of these estimates is still large, and it is necessary to enhance earthquake monitoring capability through upgrade and deployment of seismic networks with relevant infrastructures, and establishment of data centers. In addition, it is important to study local and regional seismicity, since many earth tremors has been felt recently in the country. I would like to propose the following suggestions future development.

(1) The current University seismic network, which at present is not working properly, should be upgraded so that it is possible to monitor earthquakes continuously. The upgrade plan was studied and proposed by Thaldena (2008). It is desirable to establish a real-time data transfer between these stations and GSMB National Seismic Data Center.

(2) In order to better understand seismicity in and around the country, another seismic network with sufficient number of short period seismometers is desirable.

(3) The PALK station is governed by international agencies and there are certain limitations for data usage. It is highly desirable to establish our own broadband seismic station. This would be definitely beneficial for earthquake and tsunami disaster mitigation of the nation.

ACKNOLEGEMENTS

I would like to express sincere gratitude to Ms. Nilmini Thaldena at Geological Survey and Mines Bureau of Sri Lanka for providing me data and information during the study.

REFERENCES

Fernando M.J. and Kulasinghe A.N.S., 1986, Phys. Earth planet. Inter., 44, 99-106.
Droznin, D., 1997, Operations manual for USGS, Kirtland, Albuquerque.
Kennett, B.L.N. and Engdahl, E.R., 1991, Geophys. J. Int., 105, 429-465.
Thaldena, S.N.B., 2008, Bulletin of IISEE, 42, 37-42.

Synopsis of Master Papers

*Bulletin of IISEE, **43**, 49-54, 2009*

GEOLOGICAL INDICATORS ASSOCIATED WITH PALEOEARTHQUAKES: FOCUSING ON THE GEOMORPHOLOGICAL EVIDENCES

Sule GURBOGA*
MEE07168

Supervisor: Yuichi SUGIYAMA**

ABSTRACT

In this study, I have categorized the evidences into two groups; geomorphological evidences (macro-scale) related with surficial expressions and stratigraphic evidences (micro-scale) which is obtained from trench studies. This study is focusing on the geomorphological evidences to find the location of fault, to prove the existence of large earthquakes or not in the past from the geological point of view. Geomorphological evidences which are obtained from aerial photography and geological field excursion have been done for Tachikawa Fault, Itoigawa-Shizuoka Tectonic Line Active Fault System, and Sekiya Fault to test the reliability of geomorphological features indicating fault on the surface. Some of the evidences which are topographic altitude differences between both sides of the fault, displaced landforms, drainage patterns cut and displaced by main fault, geological unit differences, and lineament analysis have been investigated by using the aerial photographs for study areas. Other evidences which are used to check the reliability of data coming from aerial photography study, detailed geological unit description to find the deformation pattern produced by fault and to determine the exact location of fault, and preparation of geological map to define the boundary of units have been obtained by geological field excursion. For three study areas, firstly aerial photography study has been done to get the general idea about areas. Secondly, field excursion has been carried out to control the data from aerial photos and to obtain new evidences by using the geological evidences. Also, to examine the recent surficial expressions, Iwate-Miyagi Earthquake area is selected and geomorphological study has been carried out to discuss the critical situation for paleoseismology.

Keywords: Paleoseismology, Paleoearthquakes, Geomorpgological evidences.

1. INTRODUCTION

Paleoseismology an interdisciplinary science between Geology and Seismology is purely geological exploration for paleoearthquakes. Paleoseismology is the study of prehistoric and historic earthquakes (Solonenko, 1973; Wallace, 1981) especially their location, timing and size (McCalpin, 1996) mainly by using the field investigations. It is deeply concerning with the identification of active faulting, amount of slip rate, rupture length, repeated time, and slip per event and assessment of magnitude of future events. The main reasons to understand the paleoearthquakes are activation of the same fault with similar behavior and the latest earthquake on the fault which are critical to judge the hazard coming from that fault (Schwartz and Coppersmith, 1984). Although it is very difficult to get information about past events implying pre-historical or historical events, paleoseismological investigations have proven all around the world to be very effective device for assessing the seismogenic potential of any given fault.

The main purpose of this study is to clearly document the relationship between geomorphological and structural behavior of the geological units to be exposed to destructive earthquakes in the past. This study is supposed to give clear explanation about geomorphological evidences to investigate the large earthquakes happened in the historical and pre-historical time.

---

*Middle East Technical University, Ankara, Turkey
** National Institute of Advanced Industrial Science and Technology (AIST), Geological Survey of Japan, Active Research Fault Center

2. METHODOLOGY

The first and significant method in paleoseismology is aerial photography interpretation. Totally, 193 aerial photos (Tachikawa Fault (98), Itoigawa-Shizuoka Tectonic Line Active Fault System (45), and Sekiya Fault (50)) have been examined. By using the aerial photos, all topographic features presenting in the study areas are carefully described and then some critical places which have sharp vertical changes on the slope, drainage displacements and displaced landforms have been signed to visit later during field study. Therefore, lineament maps have been drawn. Second method used to define geomorphological evidences is field excursion. After examining the aerial photographs, I selected some critical places which have strong evidences like sharp changes in altitude, slope and also some clear lineaments which are possible to accurate location of fault. Before visiting these areas, I prepared general geological map of the study areas by using the aerial photos. It should be prepared very detail but time is not enough to do so. After the step, I visited and tried to observe the fault plane.

In this study, I divided into two groups of indicators; the geomorphological (macro scale) and stratigraphical indicators (micro scale). The main part of this thesis is geomorphological indicators from the three examples in Japan. But stratigraphic indicators have been explained briefly.

3. STRATIGRAPHIC INDICATORS (MICRO SCALE EVIDENCES)

Stratigraphic indicators to define the paleoearthquakes are also very effective to get the actual ages of units and events. Geomorphological and stratigraphical evidences are used to describe different part of events. For example, geomorphological evidences are used to find the location of faults and all information included by surficial expression. On the other hand, stratigraphic clues can give fault plane itself by means of trenching study. For stratigraphic evidences, trenching study is an influential application.

In trenching studies, these laws are the key to define the past events or any abnormal situation. The relationship between sedimentary structures must be examined carefully. Before constructing the faulting history of any trench site, the characteristics of sedimentary units, and deformation features must be recorded in detail. In some cases, we cannot observe the fault because of lack of permanent clue. In this case, the deformation patterns of units become very important.

The classification of stratigraphic indicators can be very difficult in complex structural events. They can be divided into on fault, off fault, direct and indirect evidences according to their probability percentages for past event (Allen, 1986; Toda, 1998). These authors described some features used to identify individual paleoearthquakes obtained from trench survey under this classification. In this study, I explained all situations and divided into four groups; faulting related, depositional patterns, strong shaking and borehole and coring evidences.

In next part of the report, I have explained my study geomorphological approaches for three main faults in Japan.

4. AERIAL PHOTOGRAPHY AND FIELD EXCURSION ON STUDY AREAS (JAPAN)

I have explained aerial photography study and geological field excursion on the study areas in Japan. These are Tachikawa Fault, Itoigawa-Shizuoka Tectonic Line Active Fault System, and Sekiya Fault. Former three areas have been selected as an application area of aerial photos and field excursion methods to determine the active faults. Besides, latter one Iwate-Miyagi earthquake area has been selected to discuss the situation in which before large earthquake there is no any surficial expression. It is explained in discussion part in detail.

The reason to choose three main areas for application is that in paleoseismology there are many geomorphological indicators and it is not possible to describe many of them in only one place. To get many evidences and define them clearly, we have chosen three areas. These study areas have been explained separately.

4.1. Tachikawa Fault

First of all, I have examined 98 aerial photos totally. I have prepared two maps including lineament and drainage pattern map to define the critical area to visit during field excursion. These two maps are very important to find the location of fault. In lineament map, big vertical differences cannot be detected. For Tachikawa fault, drainage map is not useful to define the location of map. There is not any clear deformation or lateral displacement on the map.

After aerial photograph study, we have visited the place and checked the lineaments seen on the aerial photos. There are some clear displacement along highway and agricultural field (Figure 1). In this figure, black arrows are showing the part of fault which has been affected by human activity and white arrow is showing the actual displacement on the fault.

For Tachikawa fault, the data obtained from the aerial photography study to define the location of the fault has not satisfied. Without field study, it can be difficult to define the exact place of fault.

Figure 1. Location of Tachikawa fault.

4.2. Itoigawa-Shizuoka Tectonic Line (ISTL)

45 aerial photographs have been examined to obtain 3-dimensional view to identify topographical expressions. The most important part of the study for this fault, topographic altitude differences between both sides of the fault is very high and it has been detected obviously from aerial photos.

ISTL cover very large place on the centre of Honshu Island. We only focused on the central part of it. Figure 2 is showing one of the aerial photography interpretations in this fault. Topographic elevation differences between two sides of the fault and drainage pattern change are very important clues to define the location of fault. Dark color area is showing the mountainous area. After aerial photos study, I defined many place to visit during field excursion. One of the selected areas is shown on Figure 2 in circle. Close up view of the area can bee seen on Figure 3. The exact location of fault has been detected successfully.

After aerial photography study, many places selected during this study have been visited and all of them are the place illustrating the fault clearly. For paleoseismological study, next step should be trench site selection. In this study, because of the time limitation, trenching study cannot be applied.

The most important issue in Figure 3 is that older unit is located on higher elevation and younger unit is located on lower elevation and they have very sharp boundary.

Figure 2. Aerial photo interpretation of the central part of the ISTL, Shimotsuburai Fault.

Figure 3. The close up view of fault from SSE to NNW on the area signed by circle in Figure 2.

Topographic elevation differences, drainage pattern, and lineaments maps from aerial photography and geological unit boundaries, reverse location of younger and older units obtained from field study are apparently prove the location of fault.

4.3. Sekiya Fault

For this fault, I have looked at 50 aerial photos to prepare lineament map and drainage pattern map. In lineament map, the trace of the fault is signed by large vertical displacement. Western site of the fault is mountainous area and it has steep slopes, but, eastern site of the fault is almost flat-lying area represented by recent alluvial fans. By using the altitude differences and sharp slope changes between two sides of the fault, and lineament pattern, it is very easy to find the location of fault from aerial photos.

In Figure 4, location of the fault from aerial photos has been shown by white arrows. The area showing in the figure 3 is very big and it is very difficult to walk along the fault. Because of this reason, I selected some places to visit and the situation is the same in ISTL, fault location can be seen easily during the fault excursion.

I have visited not only the places which have big vertical elevation differences but also the small elevation differences on the alluvial deposits to get the idea for further trenching studies.

As a conclusion, surficial expressions elevation and slope change, drainage pattern, lineament maps from aerial photograph studies, field excursion for geological description and evidences and results obtained from past trenching studies are used to get reliable data about fault location. These criteria have been checked for each study areas, and concluded.

Figure 4. Aerial photograph of tip area from Sekiya fault.

5. DISCUSSION AND RECOMMENDATION

Paleoseismology is very important tool to detect the past events by using the geomorphological evidences. For Tachikawa fault, Itoigawa-Shizuoka Tectonic Line (ISTL), and Sekiya fault, many of the geomorphological indicators have been observed and the location of fault has been defined. Yet, it is possible not to observe any geomorphological evidences without any recent large earthquake. Iwate-Miyagi Nairiku Earthquake (13.06.2008) is very good example for such difficulties in case of no surficial expression before large event. According to USGS, the epicentral location of event is 39.122°N, 140.678°E, magnitude is Mw=6.8.

Before occurring the earthquake, there was not any clear indicator about existence of active fault here. That is the reason to select this area as a good example in which geomorphological evidences cannot be seen or used to define the fault location.

After one month, we visited the place to examine the surface rupture and surficial expression occurred after event. We walked along the central part of the surface rupture and observed the displacement very clearly. Surface rupture cut and displaced many rice fields. The amount of vertical offset is around 20 cm.

Vertical displacement range is 20-55 cm which I measured around this part of the fault. These displacements are located on rice fields so before that earthquake, the detection of fault and existence of active fault are very difficult because of the flat rice fields. During this trip, I also observed many landslides because of strong shaking. But these are not useful to define the exact location of fault.

Figure 5. Deformation of pine trees and vertical displacement on the fault.

In Figure 5, I measured max displacement 55cm. As you see the pine tree and vertical displacement, location of fault is very clear. The letters a and b are showing the position of pine trees before and after the earthquake, respectively. Location and the displacement of fault are signed by white line and arrow in the figure. This point is the clearest location of fault even before the earthquake. Small amount of displacement and lack of indicators in other parts of the fault are the reason that this place was not evaluated as an active fault location. It means that if there was not any earthquake here, the existence of active fault could not be detected. This situation is one of the important limitations of paleoseismology. If you do not have or very little surficial expression which cannot be realized during aerial photography or field survey study, it is very difficult to judge the location of fault even the existence of fault.

In such situation, there is only way to detect the existence of fault in any place that is historical documents if the nation has.

6. CONCLUSIONS

In this study, Tachikawa fault, Itoigawa-Shizuoka Tectonic Line Active Fault System, and Sekiya fault located inland part of the Japan have been studied carefully. Aerial photography study and detailed field excursion study carried out for these areas. Additionally, Iwate-Miyagi earthquake has been examined as a difficult case for paleosesimological application. For Tachikawa fault, 98 pieces of aerial photos have been examined. Because of the small slip amount of fault, a clear vertical displacement has not been observed. After preparing the lineament, drainage map and topographic altitude differences, I checked them in the field. During the field study, the fault location can be seen with minor amount of displacement same with the results coming from aerial photos. The reason for the small amount of displacement is that this place has a dense construction and fault trace on the surface destroyed by them, but, still there is a flexural surface rupture in some parts. Combination of aerial photography results and field excursion study, the existence of fault and past event which has been produced by this fault have been detected. For central part of the Itoigawa-Shizuoka Tectonic Line Active Fault System, 45 aerial photos are examined. This part of the fault is strike-slip with thrust component. Because of the thrusting, topographic altitude differences between two sides of the fault very clear. Drainage pattern is signing the location of fault properly without any doubt. Therefore, lineament map of the aerial photos is also showing the same location of fault getting from other indicators. During field study, triangular facets, river offsets, sharp boundary between two different units and their order (older is upper, younger is lower elevation) are clearly indicating the existence of fault and past earthquake here. For Sekiya fault, 50 aerial photos have been studied. This fault has thrust character dominantly. That is why I observed a very large vertical displacement between up-thrown and down-thrown blocks of the fault. Depending on this feature, lineament map of the area is obviously indicating the location and existence of fault.

Lineaments, and drainage maps, altitude differences on both sides of the faults, geological unit description and boundary between them have been defined to check closely during the aerial photography and field excursion. Aerial photography study is the first step to look the field as a general point of view. Geological field study is the second step to confess the information coming from aerial photographs. In addition, by using the geomorphological evidences, I only described the existence of fault and past large events. I cannot say anything about how many earthquakes occurred in the past and their times. For these things, trenching study is needed. Furthermore, detailed geological mapping is very important for next step which is trenching. For want of time, trench study could not be applied in this study. One of the most important data obtained from trenching investigation is the stratigraphical indicators for paleoearthquakes. Besides, I have summarized the information about necessary tasks and useful Stratigraphical Indicators for my studies in future.

AKNOWLEDGEMENT

It is a highly rewarding practice gained by working with Dr. Shinji Toda, to whom I am indebted for his experiences and discussion during the preparation of this study. I would like to say thanks to Dr. Yokoi for his invaluable comments and corrections to get the last version of the thesis.

REFERENCES

Allen et al., 1984. Active Tectonics, Studies in Geophysics. National Academy Press, p. 266.
McCalpin J.P., 1996. Paleoseismology, Academic Press, 588p.
Schwartz and Coppersmith, 1984. Journal of Geophysical Research, Vol. 89, No. B7, Pages 5681-5698.
Solonenko, V.P., 1973. Izvestia, Solid Earth, v.9, p.3-16 (in Russian).
Toda, S., 1998. PhD Thesis, Tohoku University, 205 pages.
Wallace, R. E., 1981. Earthquake Prediction, Maurice Ewing Series 4: American Geophysical Union, p. 209-216.

Synopses of Master Papers   *Bulletin of IISEE, **43**, 55-60, 2009*

APPLICATION OF RECEIVER FUNCTION TECHNIQUE TO WESTERN TURKEY

Timur TEZEL[*]   Supervisor: Takuo SHIBUTANI[**]
MEE07169

ABSTRACT

In this study I tried to determine the shear wave velocity structure in the crust and uppermost upper mantle using receiver function technique. For this purpose I selected teleseismic earthquakes in the epicentral distances between 30 and 90 degrees, magnitudes of which are greater than 6.3. I collected waveform data recorded at Turkey's broadband seismic stations between 2006 and 2008. I used radial receiver functions which were calculated using extended time multitaper method to determine the crustal and uppermost upper mantle structure. My study consists of three steps: receiver function (RF) calculation, RF image and RF inversion. Because of the data availability, I studied data of 33 stations in western Turkey which belong to the General Directorate of Disaster Affairs Earthquake Research Department and Bogazici University Kandilli Observatory. First, I calculated radial receiver functions for each station. In the next step, I made two RF images using the RFs from stations along two lines: the one is running from the northeast of Bozcaada to Konya city and the other is running from the north of Izmir to the northeast corner of Rhodes island. In the third step, I applied genetic algorithm inversion method to determine the shear wave velocity structure beneath six seismic stations.

The results show that the Moho depth changes from region to region; we observed that the Moho depth is around 35 km and 25 km beneath the former and latter profiles, respectively. The results of receiver function inversion are consistent with these RF images. Shear wave velocities were estimated to be in the range between 3.5 and 3.9 km/s for the lower crust and between 4.0 and 4.6 km/s for the uppermost upper mantle, respectively.

Keywords: Receiver Function, Crust, Teleseismic Wave, Turkey.

INTRODUCTION

Turkey has experienced many natural disasters which have caused serious casualty, collapse of buildings, economical losses up to now. Among these disasters, earthquakes are the first to be concerned rather than land slides, floods, rock falls, avalanches, and droughts. In Turkey, earthquakes occur in the upper crust generally, and one of the important issues is to determine discontinuities and velocity changes in the crust and also to determine the transition to the Moho. There are some seismological studies to find a velocity model using travel time tomography, surface wave group velocity inversion, reflection profiling, and receiver function method etc. Determination of the discontinuities and especially shear wave velocity structure has an important role in the planning of the urban areas and cities. In this study, we applied receiver function method to determination of shear wave velocity structures beneath the seismic stations in Turkey.

---

[*]General Directorate of Disaster Affairs Earthquake Research Department, Ankara, Turkey.
[**]Assoc. Professor, Disaster Prevention Center, Research Center of Earthquake Prediction, Kyoto University, Japan.

DATA

I used teleseismic waveform data for 50 earthquakes recorded by General Directorate of Disaster Affairs Earthquake Research Department (here after ERD) and Boğaziçi University Kandilli Observatory (KOERI) broadband stations. I selected 33 broadband stations which are located in western Turkey. Selection criteria were as follows: magnitude of earthquakes should be greater than 6.3 and epicentral distance should be between $30°$ and $90°$.

Data were obtained from KOERI via internet (http://barbar.koeri.boun.edu.tr) and from ERD via CD-R.

THEORY AND METHODOLOGY

The Receiver Function technique is used for determination of crustal structure beneath seismic stations. Teleseismic records $(30° \leq \Delta \leq 90°)$ are used in this method. Teleseismic P waveforms contain information related to source, propagation path and local structure beneath the recording station. The method uses the coda part of the P wave which includes converted phases and reverberations generated at discontinuities beneath each station and convolved with source function and instrumental impulse responses. If we eliminate the source and instrumental effects from waveforms, they provide information about the local velocity structure under the seismic station. In Figure 1, we can see the simple ray diagram for the incident wave and its converted and reverberated phases.

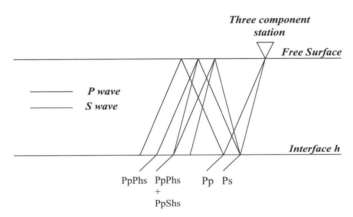

Figure 1. Simplified ray diagram identifying the major P to S converted phases which comprise the receiver function for a single layer over a half-space under the seismic station (modified from Ammon(1990)).

A teleseismic P wave arrives at the recording station with both a constant and relatively large horizontal phase velocity (15-25 km/s). This phase velocity justifies a plane wave and simplifies the study of the resulting ground motion (Ammon, 1990).

Receiver Function Calculation

Determination of the velocity structure of the crust and upper mantle beneath a single seismic station can be done by using teleseismic receiver functions. The receiver function analysis uses the converted phases and multiples recorded on the horizontal seismograms (e.g. Burdick and Langston, 1977; Langston 1977, 1979; Owens *et al.*, 1984; Ammon, 1991).

$$D_V(t) = I(t) * S(t) * E_V(t)$$
$$D_R(t) = I(t) * S(t) * E_R(t)$$
$$D_T(t) = I(t) * S(t) * E_T(t) \tag{1}$$

Here, V, R, T shows vertical, radial and tangential components respectively. Also, $I(t)$ is impulse response of the recording instrument, $S(t)$ is the seismic source function, $E_V(t)$, $E_R(t)$, $E_T(t)$ are the vertical, radial and tangential impulse response of the earth structure. And asterisk (*) shows the convolution operator.

As earth structure beneath a station will produce phase conversions of the P to S type, horizontal components of ground motion will be different from the vertical component. $D_V(t)$ contains the factors which we wish to remove from observed seismograms, so isolating $E_R(t)$ and $E_T(t)$ can be accomplished by deconvolving $D_V(t)$ from $D_R(t)$ and $D_T(t)$. In receiver function method, there are several stabilization methods; they are the water level method, the multitaper method, the extended time multitaper method. In this study I used the extended time multitaper method which was improved by Shibutani et $al.$ (2008) based on Park and Levin (2000) and Helffrich (2006). In this method, they used the three lowest-order 4π prolate eigentapers of 50 sec duration. They summed the multitapers with 75 % window overlap. The multitapers smoothly are connected at the junctions, and the resultant taper has a flat level. If we use the flat part for windowing the P onset and the P coda, we can estimate the relative amplitudes of receiver functions (Shibutani et $al.$, 2008). As a result, the frequency-domain receiver function is defined by

$$E_R(\omega) = \frac{\sum\limits_{k=0}^{K-1} H^{(k)}(\omega)\tilde{U}^k(\omega)}{\sum\limits_{k=0}^{K-1} U^{(k)}(\omega)\tilde{U}^{(k)}(\omega) + \left(\dfrac{J_S}{J_N}\right)^2 \sum\limits_{k=0}^{K-1} N^{(k)}(\omega)\tilde{N}^k(\omega)} G(\omega) \tag{2}$$

where $H^{(k)}(\omega)$ and $U^{(k)}(\omega)$ denotes the Fourier spectrum of the radial or transverse and vertical components of the waveform data with the kth prolate eigentaper, and $N^{(k)}(\omega)$ is the Fourier spectrum of presignal noise of the vertical component waveform. J_S and J_N are the numbers of the multitapers used for the signal and presignal. The second term in the right-hand side of equation (2) is a Gaussian high cut filter in which a controls the the corner frequency. In this study, I set a to 2, and then the corner frequency becomes 0.3 Hz. I showed calculated radial receiver functions with this method for two stations (GDZ and KDHN) in Fig. 2. In these figures, we can see the difference between waveforms of receiver functions related with backazimuth. It reflects the complexity beneath the stations and the lateral change of earth structure.

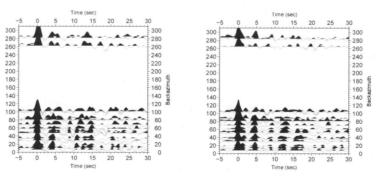

Figure 2. Calculated receiver functions for the GDZ and KDHN stations are shown on the left and right panel, respectively. Receiver functions are ordered by backazimuth and positive amplitudes are colored by red.

Receiver Function Image

The peaks and troughs of receiver functions correspond to boundaries of the S-wave velocity structure, and the time axis of the receiver functions can be converted to the depth axis using a 1-D velocity model. Delay time between converted Ps phase and direct Pp phase is given by the below equation,

$$T_{ps}(Z) = \int_{z=0}^{z=Z}\left(\sqrt{\frac{1}{\beta(z)^2} - p^2} - \sqrt{\frac{1}{\alpha(z)^2} - p^2}\right)dz \qquad (3)$$

where α and β are P and S wave velocities as a function of depth z respectively, Z indicates the depth of the interfaces, and p is the ray parameter. The converted receiver functions can be represented by a bending ray with the ray parameter and a backazimuth. I projected the rays into 1 km by 1 km cells. I used four 1-D velocity models: the three models W1, W2, and W3 are based on Tezel *et al.* (2007) and the fourth model is AK135 (Kennett *et al.* 1995). Figure 3 shows the 1-D velocity models (W1, W2, W3 and AK135) and two lines: the one is running from the northeast of Bozcaada to Konya city and the other is running from the north of Izmir to the northeast corner of Rhodes island along which RF images are processed. Figure 4 shows the images obtained by this method. In the figures, the positive amplitudes of the receiver functions are indicated by the red color and indicate velocity discontinuities at the top of high velocity layers. In Figure 4, 2-D RF images reflect the relation between earth structure and 1-D velocity models W1, W2, W3 and AK135 respectively.

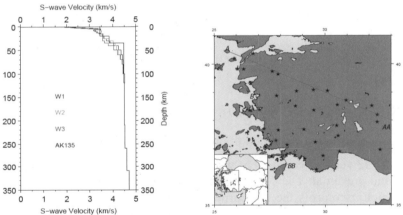

Figure 3. Input 1-D velocity models based on Tezel *et al.* (2007) and Kennett *et al.* (1995) and selected image lines (A-AA and B-BB).

Receiver Function Inversion

In previous studies, a linearization procedure was used to invert the receiver function which requires the initial model to be close to the true velocity structure. Ammon (1990) showed that the final models were dependent on the initial models. For this reason in this study I used the genetic algorithm (GA) (Shibutani *et al.* 1996).

In this study, I applied this inversion technique to radial receiver functions of the six broadband stations. For this purpose radial receiver functions were stacked according to the backazimuths and RF waveform similarities for each station. I tried to model the crust and uppermost mantle down to 50 km with six major layers: a sediment layer, basement layer, upper crust, middle crust, lower crust and uppermost mantle. The model parameters in each layer are the thickness, the S wave velocity. The velocity ratio between P and S waves (Vp/Vs) and the density in the sediment and basement layers are also model parameters. For each model parameter, upper and lower bounds and 2^n

possible values are specified. The size of the model space to be searched is $2^{46} \sim 7.04 \times 10^{13}$. Figure 5 shows the Moho depth estimated for these stations.

DISCUSSIONS AND CONCLUSIONS

In this study I applied RF (receiver function) technique to determine the velocity discontinuities such as Moho etc. Moreover, I applied genetic algorithm using receiver functions to determine the shear wave velocity structure beneath six seismic stations. I used only radial receiver functions (I calculated the tangential receiver functions but not used them in the interpretation). After calculation of receiver functions, I employed 2-D image technique which converts the time axis to depth axis using 1-D velocity models and shows the amplitudes with colors. Red (blue) color shows positive (negative) amplitude which indicates the increase (decrease) in the velocity. Along A-AA and B-BB lines I applied this method and some velocity discontinuities were detected and indicated by red dotted lines (Fig. 4).

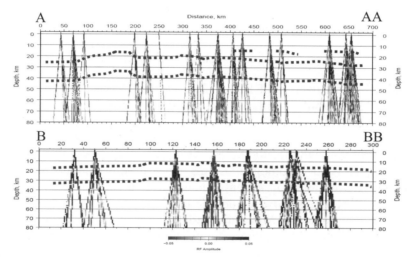

Figure 4. RF images for the A-AA and B-BB lines using W3 1-D velocity model.

We interpreted that Moho should lie between 30 and 40 km for line A-AA and probably around 35 km. For some stations we can see another signature around the 15 km, which was interpreted as Conrad discontinuity. In line B-BB Moho discontinuity seems placed between 20 and 30 km depth and probably around 25 km.

After RF image process I employed the genetic algorithm to determine the shear wave velocity model beneath seismic stations. I selected six broadband stations (BALB, GDZ, ALT, KDHN, LADK, KONT) that have more data and show good RF results. I stacked the RFs according to the backazimuths and then applied inversion process for each station.

Synthetic receiver functions generally show good fitting with the observed waveforms. Shear wave velocity models which were derived from inversion indicate that general character of crust – mantle boundary in the region. The depth of the boundary for each station was determined. Shear wave velocities near the surface varied between 1.5 and 3.5 km/s and at the mid – crustal depth range between 3 and 4 km/s. Sub Moho velocities are between 4 and 4.6 km/s. The results for some groups (made based on backazimuths) for the stations KONT, ALT and GDZ indicate low velocity zone between 25 and 35 km.

Figure 5 show the estimated Moho depths for six stations. These velocity values are generally similar to those of previous studies, while we can not observe low velocity layer around 10 km that was mentioned by previous studies.

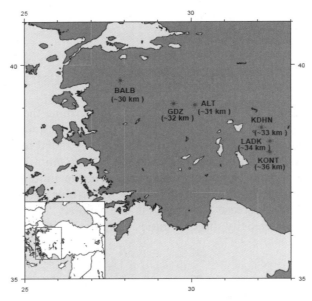

Figure 5. Moho depths calculated from inverison

Results of this study with RF image and RF inversion show the Moho thinner in the west of study area than in the east part. These results similar to the previous studies which were mentioned the Moho depth changes from 30 km to 38 km from the west to the east of Turkey. Moreover, other stations which were used for the receiver function show complex receiver waveforms, which reflect the complex tectonic structure around the study area. Also, difference between A-AA and B-BB profiles suggest that the study area is not a homogenous area and that deformation is different in each area. In these RF images Moho discontinuity has been seen as a wide band caused by the Gaussian filter used the mean value for the Moho depth considering the error range (i.e. ± 5 km). This study is a preliminary attempt to determine the Moho discontinuity based on RF image and will contribute to construction of a reliable reference velocity model in Turkey.

AKNOWLEDGEMENT

I would like to express my sincere gratitude to my adviser Dr. T. Hara for his suggestions and contributions to my study.

REFERENCES

Ammon, C.J., 1990, J. Geophy. Res., 95, 15303-15318.
Ammon, C.J., 1991, Bull. Seism. Soc. Am., 81, 2504-2510.
Burdick, L.J., and Langston, A.C., 1977, Bull. Seism. Soc. Am., 67, 677-691.
Helffrich, G., 2006, Bull. Seism. Soc. Am., 96, 344-347.
Kennett B.L.N., Engdahl E.R., and Buland, R., 1995, Geophys. J. Int., 122, 108-124.
Langston, C.A., 1977, Bull. Seism. Soc. Am., 67, 1029-1050.
Langston, C. A., 1979, J.Geophys.Res., 84, 4749-4762.
Owens, T. J., Zandt, G., and Taylor, S.R., 1984, J.Geophys. Res., 89, 7783-7795.
Park, J., and Levin, V., 2000, Bull. Seismo. Soc. Am., 90, 1517-1520.
Shibutani, T., Sambridge, M., and Kennett, B., 1996, Geophy. Res. Let., 23, 1829-1832.
Shibutani, T., Ueno, T., and Hirahara, K., 2008, Bull. Seismo. Soc. Am., 98, 812-816.
Tezel, T., Erduran, M., and Alptekin, Ö., 2007, Annals of Geophysics, Vol.50, No.2, pp. 177-190.

DAMAGE DETECTION OF REINFORCED CONCRETE STRUCTURE BASED ON LIMITED STRONG MOTION RECORDS

Gong Maosheng[*]   **Supervisor: Toshihide Kashima**[**]
MEE07151

ABSTRACT

The author presents one simplified technique and methodology for structural parameter identification and global damage detection from limited number of strong motion records. The time-invariant and time-varying parameters of structure are identified by using the off-line system identification method ARX and the on-line system identification method RARX respectively. The presence of damage is detected primarily by checking the changes of parameters during the earthquakes. Then, one procedure is introduced to convert MDOF structural system to equivalent SDOF system. The integral damage level of the equivalent SDOF system is detected by using the ductility factor and damage index. The damage index of the equivalent SDOF system can be considered as the integral damage level indicator of the MDOF system. The method is applied to Hachinohe City Hall building which was lightly damage in the Sanriku-oki Earthquake on Dec. 28, 1994. The parameters of the building are identified and the damage level is detected from the strong motion records. The results show that the method has with high efficiency, validity, applicability and practicability. The method can be implemented to assess the structural damage level and to judge if the damage is repairable or not after a big earthquake. Furthermore, the method can be used to predict the failure of a whole structure and to evaluate the performance in earthquake. The presented technique and method can be considered as a significant aid for making structural retrofitting decision after devastating earthquake.

Keywords: Strong motion record, Damage detection, System identification.

INTRODUCTION

The structural damage detection aims to detect, localize and classify damage of structures and also to predict and assess the safety and remaining service life of structures. Many kinds of techniques and methods have been developed and applied for the purpose in the past decades (Doebling, 1996; Sohn, 2003; Farrar, 2007). In fact, the damage is not meaningful without comparison between undamaged state and damaged state. That means the structural parameters before and after damage are critical information for damage detection. Determining the dynamic properties of a structural system from the response data is well known as system identification, and many kinds of methods have also been developed. Generally speaking, the damage detection is related to or based on the structural parameter identification at a certain extent. To monitor the structural response and performance during earthquakes, strong motion seismographs are installed in many buildings. The response data recorded by the strong motion observation system during earthquakes can be used to determine the structural parameters and damage state (Loh, et al, 1996, 2000). It is very important to judge damage level after earthquake especially for the buildings slightly or moderately damaged, but sometimes it is difficult to check the damage by visual inspection since the structural members are covered with finishing materials. In such case, it has remarkable meaning to detect the damage and determine the damage level by using the strong motion records. Such kind of post-earthquake damage assessment and

---

[*]Institute of Engineering Mechanics, China Earthquake Administration, Harbin, China.
[**]International Institute of Seismology and Earthquake Engineering, Building Research Institute, Japan.

damage detection has significant life-safety implication, because it helps to determine whether the building is safe enough for reoccupation. Furthermore, it has great meaning to judge if the building can be repaired and reoccupied from the economy issues. According to the analysis stated above, the author presents simplified methods to identify the parameters and diagnose the damage level after huge earthquake from the limited strong motion records.

METHODOLOGY

System Identification

Two system identification methods are used to obtain the structural parameters. One is the off-line ARX (Auto-Regression with eXogenous) method for time-invariant parameters, the other is RARX (Recursive ARX) method for the time-varying parameters. The ARX model can be denoted as:

$$y(t) = \varphi^{\mathrm{T}}(t)\theta + e(t) \tag{1}$$

where θ is the vector including unknown system parameters; $\varphi(t)$ is the vector including input and output samples. The solution of Eq. (1) can be estimated by the following Eq. (2) (Ljung, 1999):

$$\hat{\theta}_N = \left[\sum_{t=1}^{N} \varphi(t)\varphi^{\mathrm{T}}(t) \right]^{-1} \sum_{t=1}^{N} \varphi(t) y(t) \tag{2}$$

The modal parameters can be calculated once the system parameters are obtained as shown below:

$$\omega_r = \frac{1}{\Delta t} \sqrt{\ln Z_r \ln Z_r^*}, \quad \xi_r = \frac{-\ln(Z_r Z_r^*)}{2\sqrt{\ln Z_r \ln Z_r^*}} \tag{3}$$

where (Z_r, Z_r^*) is the r-th discrete complex conjugate eigenvalue pair and Δt is the sampling period. ω_r and ξ_r are the r-order modal frequency and damping ratio respectively.

To track changes of structural parameters during the vibration, RARX method is used to identify the time-varying parameters. RARX is a system algorithm to estimate recursively parameters of ARX model in Eq. (1) by using RLS method. The solution can be estimated as following (Ljung, 1999):

$$\hat{\theta}(t) = \hat{\theta}(t-1) + P(t)\varphi(t)\varepsilon(t) \tag{4}$$

$$\varepsilon(t) = y(t) - \varphi^{\mathrm{T}}(t)\hat{\theta}(t-1) \tag{5}$$

$$P(t) = \frac{1}{\lambda}\left(P(t-1) - \frac{P(t-1)\varphi(t)\varphi^T(t)P(t-1)}{\lambda + \varphi^T(t)P(t-1)\varphi(t)} \right) \tag{6}$$

Then the equivalent stiffness K_{eq} can be calculated by using the following Eq. (7):

$$K_{eq} = \frac{4\pi^2 M}{T^2} \tag{7}$$

where, M is the total mass of the structure or system; T is the fundamental period of the structure.

MDOF System to Equivalent SDOF System

To estimate global damage index of structures under the excitation of earthquake based on the limited

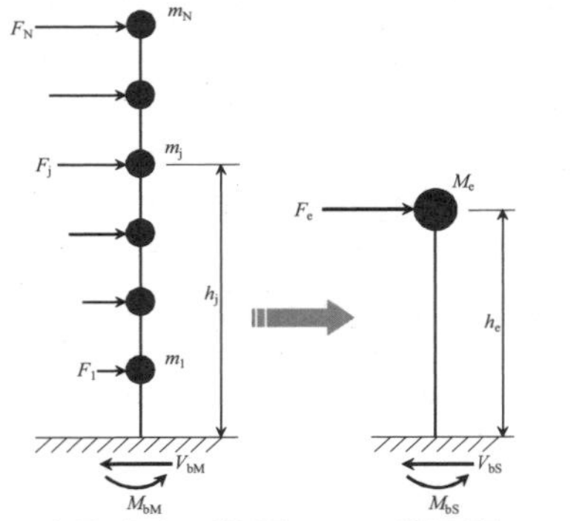

Figure 1. Equivalent SDOF system of MDOF system

strong motion response records, the MDOF system is converted to equivalent SDOF system. The total mass M is assumed as the equivalent mass M_e of the SDOF system as shown in Eq. (8):

$$M_e = \sum_{i=1}^{N} m_i \tag{8}$$

where m_i is the mass of the i-th story of MDOF system. Then the distribution of the earthquake shear force of the MDOF system is assumed as the following Eq. (9) (AIJ code):

$$A_i = 1 + \left(\frac{1}{\sqrt{\alpha_i}} - \alpha_i \right)\frac{2T}{1+3T} \tag{9}$$

where the α_i is the normalized weight of the i-th story, which is calculated as the weight above i-th story divided by the weight above ground. T is the natural period. If the response data on roof is recorded, the earthquake force on the i-th story can be estimated by the following Eq. (10):

$$F_N = m_N a_{N\max}, \quad F_i = \frac{A_i}{A_N} F_N \qquad i = 1,2 \cdots N \tag{10}$$

where m_N is the mass of top story and $a_{N\max}$ is the maximum acceleration on the top. From $V_{bS} = V_{bM}$, the $a_{S\max}$ of the SDOF system under the same excitation can be estimated by Eq. (11):

$$a_{S\max} = F_e / M_e = \sum_{i=1}^{N} F_i / M_e \tag{11}$$

Ductility Factor and Damage Index Calculation

The structural damage can be described by the ductility factor and hysteretic energy dispersed by the building in earthquake. Herein, the ductility factor is estimated from the nonlinear response spectra for damage detection purpose. The concept is if we have the properties of the SDOF system, such as period, damping and ductility factor, we can find out the maximum response from the response spectra. On the contrary, the ductility factor can be determined from the constant ductility nonlinear response spectra if the maximum response of the SDOF is known.

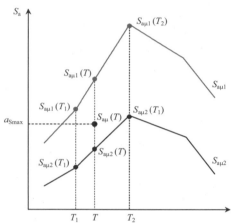

Figure 2. Interpolation relationship

The relationship between the equal ductility response spectra and maximum response of the SDOF system can be expatiated by Figure 2 from which the ductility factor can be obtained if the response spectra $S_{a\mu1}$ and $S_{a\mu2}$ of the strong motion record at base are known. From Figure 2, we can have the following the calculations:

$$S_{a\mu1}(T) = S_{a\mu1}(T_1) + \frac{T - T_1}{T_2 - T_1}(S_{a\mu1}(T_2) - S_{a\mu1}(T_1)) \tag{12}$$

$$S_{a\mu2}(T) = S_{a\mu2}(T_1) + \frac{T - T_1}{T_2 - T_1}(S_{a\mu2}(T_2) - S_{a\mu2}(T_1)) \tag{13}$$

The ductility factor of the SDOF system can be approximately estimated by Eq. (14) from $a_{S\max}$.

$$\mu = \mu_1 + \frac{S_{a\mu1}(T) - S_{a\mu}(T)}{S_{a\mu1}(T) - S_{a\mu2}(T)}(\mu_2 - \mu_1) = \mu_1 + \frac{S_{a\mu1}(T) - a_{S\max}}{S_{a\mu1}(T) - S_{a\mu2}(T)}(\mu_2 - \mu_1) \tag{14}$$

A hybrid damage index of combination of ductility and hysteretic energy is adopted to predict integral damage level and the damage index is calculated by using the Bispec software (Hachem, 2002, 2008).

PARAMETER IDENTIFICATION AND DAMAGE DETECTION

Hachinohe City Hall Building

The Hachinohe City Hall building (Figure 3) is used as the example to determine the parameters and damage level from the strong motion records. The building is a 5-story with one story basement, and RC type structure is adopted for bearing system. The building was damaged by Sanriku-oki Earthquake on Dec. 28, 1994 (denoted by EQ2). The response data were obtained in the earthquake and its aftershock (denote by EQ3). Before EQ2, the response data was recorded in one small earthquake (denoted by EQ1) on Oct. 9, 1994.

Figure 3. Hachinohe City Hall (Photograph by T. Kashima)

Time-invariant Parameters

Just for example, the time-invariant parameters in 164° direction identified from the data in EQ1 are shown in Figure 4. All the identification results are shown in Table 1 in which the parameters in both directions are also compared. From Table 1 we can see, the structural natural frequency and equivalent stiffness decreased but the damping ratios increased after the EQ2. The reduction degree of the parameters is shown in Figure 5. It can be concluded that the building was damaged by the EQ2.

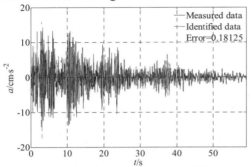
(a) Comparison of measured and identified data

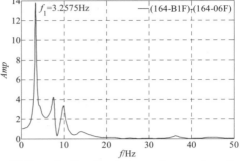
(b) Transfer function between base and roof

Figure 4. ARX Identification results from EQ1 data

Time-varying Parameters

The presence of damage can be detected by comparison of the parameters in 2 small events before and after the big earthquake. However, we don't know when the damage occurred and how is the damage level. Even we don't know damage occurred during the big earthquake if we don't have small earthquake response data before the big earthquake for comparison. For damage detection by only using strong motion records from one earthquake, the on-line RARX method is used to detect the accurate damage time and the decline trend during the earthquake. The time-varying parameters of 254° direction in EQ2 are shown in Figure 6. From which we can see the natural frequency decreased gradually until arrived at the lowest level, then, it was a little bit recovered at the tail part of the strong motion records but not reach the original value. That means the building was definitely damaged by the earthquake. The decrease of the frequency indicates the damage process during the big earthquake.

Furthermore, it can be concluded that the building started to be damaged at around 20s and stopped at around 38s from the variation of the parameters. From the above analysis, we can know that by using on-line RARX method, the damage can be detected by only using strong motion

Table 1. Identified parameters from data in the 3 earthquakes

Earthquake	f (Hz)		ξ (%)		K_{eq}(10^9N/m)	
	254°	164°	254°	164°	254°	164°
(a) EQ1	3.25	3.26	2.72	4.87	4.11	4.14
(b) EQ2	2.64	2.78	3.83	6.26	2.71	3.01
(c) EQ3	2.66	2.78	2.96	4.88	2.75	3.01
(b-a)/a(%)	-18.77	-14.72	+40.81	+28.54	-34.06	-27.29
(c-a)/a(%)	-18.15	-14.72	+8.82	+0.21	-33.09	-27.29

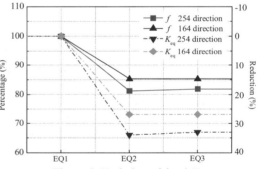

Figure 5. Variation of f and K_{eq}

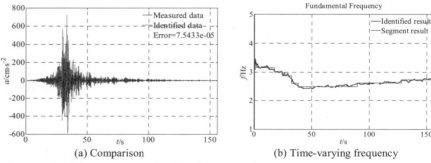

(a) Comparison (b) Time-varying frequency

Figure 6. RARX Identification results from EQ2 data (254°)

records from one earthquake. The RARX method has good qualities to detect the changes of the parameters in big earthquakes, and the parameters of the original state (undamaged) of the structure can also be determined for comparison purpose. It can be concluded the presence of damage can also be detected primarily by only using strong motion data from one earthquake.

Ductility Factor Estimation and Damage Detection

The strong motion data in 164° direction is used as the example to show the ductility factor estimation and damage detection. The building is converted to equivalent SDOF system as shown in Table 2 from which we can see the natural period is 0.3067s which is from the identification results and the peak value of the acceleration of the SDOF system is 586.44cm/s² which is used to calculate the ductility factor from the nonlinear equal ductility response spectra, and the equivalent height is 13.5m.

Table 2. Parameters of the equivalent SDOF system for Hachinohe City Hall (164°)

story	M_i(t)	$\sum_{j=i}^{N} M_j$ (t)	α_i	T(s)	A_i	PA(cm/s²)	F_i(kN)	h_i(m)
1	2141	9862	1.00	0.3067	1.000	/	8.78E+03	4.50
2	2488	7721	0.78	0.3067	1.111	/	9.76E+03	3.95
3	2070	5233	0.53	0.3067	1.269	/	1.11E+04	3.95
4	1572	3163	0.32	0.3067	1.462	/	1.28E+04	3.95
5	1591	1591	0.16	0.3067	1.744	962.60	1.53E+04	3.95
SDOF	9862	/	/	0.3067	/	586.44	5.78E+04	13.5

The nonlinear response spectra of the record at the base of the building with different ductility factors and with 5% damping ratio are shown in the Figure 7. The damping factor is adopted from the identification results from EQ1 in Table 1. According to the T=0.3067s and the maximum acceleration PA=586.44cm/s² of the SDOF system, the ductility factor can be calculated from the equal ductility spectra by using the interpolation method as shown in Table 3 from which we can know the ductility factor of the equivalent SDOF system of Hachinohe City Hall is 1.85. The ductility factor can be considered as the average ductility level of the building in Sanriku-oki Earthquake on Dec. 28, 1994.

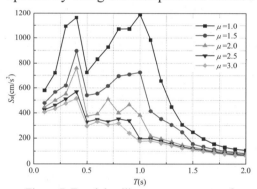

Figure 7. Equal ductility response spectra of base record 164° ($\xi = 5\%$)

Table 3. Ductility factor estimation in 164° direction

	T_1	T_2	T	μ
T	0.30	0.40	0.3067	/
$S_{a\mu 1}$	617.57	896.76	636.41	1.50
$S_{a\mu 2}$	550.17	759.24	564.28	2.00
$S_{a\mu}$	/	/	586.44	**1.85**

Table 4. Damage index in 164° direction

	T_1	T_2	T	μ
T	0.30	0.40	0.3067	/
$DI_{\mu 1}$	0.0962	0.0555	0.093	1.50
$DI_{\mu 2}$	0.1588	0.111	0.156	2.00
DI_{μ}	/	/	*0.137*	**1.85**

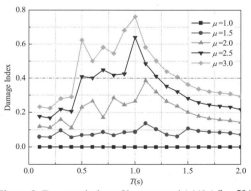

Figure 8. Damage index of base record 164° ($\xi = 5\%$)

The damage index spectra calculated from the base record (164°) are shown in Figure 8 from which the damage index of the system can be estimated according to the fundamental period T, damping ratio ξ and the ductility factor μ obtained in Table 3. The calculation result is shown in Table 4 from which we can see the damage index of equivalent SDOF system is 0.137. By using the same procedure, the damaged index in 254° direction is 0.083. Considering the real circumstance, for the same building, the maximum damage index in both directions should be adopted to describe the damage level. From this concept, the damage index of the building is 0.137. According to study of Valles, et al (1996), the damage index 0.137 means the damage of the building is slight and the damage is repairable for reoccupation after the earthquake.

By using same procedure and methodology, the damage of one 9-story RC building, Department of the Architecture and Building Science of Tohoku University which was damaged in Miyagi-ken Earthquake (Sep. 15, 1998), is detected. It is found that the damage index is 0.282 in the earthquake, which means the building is moderately damaged by the earthquake and the damage is repairable.

From the example, it can be concluded that the structural damage caused by the earthquake can be detected and assessed only by using the strong motion records on the roof and at the bottom of the structure through implementing the procedure and methodology presented in the study.

CONCLUSIONS

The author presented one method to identify parameters and detect damage level of RC buildings by using strong motion records at the top and the bottom. The parameters of Hachinohe City Hall building are identified and damage level is detected. Some conclusions are summarized as follows:

    1) The time-invariant and time-varying structural parameters including modal frequency, damping ratio and equivalent stiffness can be determined from the strong motion records by using the off-line ARX and on-line RARX system identification models. The changes of the parameters, such as decreasing of the frequency and reduction of the equivalent stiffness, can be used to find out the presence of damage with the structure.

    2) The MDOF structural system can be converted to equivalent SDOF system for damage detection purpose. The ductility factor and damage index can be obtained by using the interpolation method from the constant ductility nonlinear response spectra and damage index spectra.

    3) The ductility factor of the equivalent SDOF system can be considered as the average ductility level of structure, and damage index of the equivalent SDOF system can be considered as integral damage level indicator of the building.

    4) The damage index of structure in the study is defined as the global damage indicator and can be used to predict damage state or the failure of a whole structure after earthquake.

    5) The method presented in the study can be implemented for structural performance evaluation and for decision making of repair and continuous use of the instrumented buildings after devastating earthquakes.

ACKNOWLEDGEMENT

I would like to express my sincere and heartfelt gratitude to Dr. Taiki Saito for his continuous support, valuable suggestion and guidance during my study.

REFERENCES

Doebling, S. W., et al, 1996, Los Alamos National Laboratory report, LA-13070-MS.
Farrar, C. R., and Lieven, N. A. J., 2007, Phil. Trans. R. Soc. A 365, 623-632.
Hachem, M. M., 2002, Doctoral Thesis, University of California, Berkeley, USA.
Hachem, M. M., 2008, http://www. ce.berkeley.edu/~hachem /bispec/index.html.
Ljung, L., 1999, PTR Prentice Hall, Upper Saddle River, N.J.
Loh, C. H., and Lin, H. M., 1996, Earthquake Engineering and Structural Dynamics, 25, 269-290.
Loh, C. H., Lin, C. Y., and Huang, C. C., 2000, Journal of Engineering Mechanics, 126(7), 693-703.
Sohn, H., et al., 2003, Los Alamos National Laboratory Report, LA-13976-MS.
Valles, R. E., Reinhorn, A. M., Kunnath, S. K., et al., 1996, NCEER-96-0010.

SEISMIC LIMIT STATES AND RESISTANT MECHANISM OF THE SOIL-CEMENT BRICKS CONFINED MASONRY

Nicolas GUEVARA[1]   **Supervisor: Hiroshi FUKUYAMA**[2]
MEE07156

ABSTRACT

Performance based philosophy is proposed to verify the seismic design of the soil-cement confined masonry. Experimental wall data considering in-plane and out-of-plane behaviors have been used to propose three seismic limit states: Serviceability, Reparability and Safety. Besides that, two design seismic forces, one given in the current special normative for designing and construction of housing and another taken from recorded earthquakes, have been applied to the basic constructed house modulus to verify the safety as well as to verify its seismic design using current normative where the soil cement is not considered as infill element.

Keywords: Performance based, soil-cement brick, confined masonry

INTRODUCTION

Burnt clay bricks have been traditionally used in El Salvador as unreinforced masonry and as infill element in confined masonry. However, its use has been pointed out as harmful for the environment. The crystallization process through firing the bricks using firewood has been devastating the forest in the country. To reduce this growing problem, bricks made of soil and cement mixture have been currently used as infill element in confined masonry instead of clay bricks. Initially, research studies were focused on finding out the best soil-cement proportion without study the seismic behavior of the masonry system. Due to the earthquakes in January and February of 2001, one cooperation project started, sponsored by the Japanese International Cooperation Agency (JICA). As part of this project "Enhancement of the technology for the construction and dissemination of popular earthquake resistant housing", known as Taishin Project, which has been already executed in its first phase; the soil-cement bricks system has been studied recently under the objective of to find out its mechanical properties. Besides that, the necessity of upgrading the design philosophy, the Performance Based Design has been currently included as target for second phase of the mentioned project. The experimental research on the soil-cement has been already carried out, covering differences stages since the survey of soil around the country until decide the best soil-cement mix, construction of walls and experimental test, which is used to reach the objectives of this study. Therefore, the present study is devoted to propose performance levels or seismic limit states like Serviceability, Reparability and Safety, by using the experimental wall data obtained during the research study on soil-cement. Since the Collapse of masonry houses should be prevented, two seismic design forces are applied to the basic house modulus, one demand is taken from the recommendations given by

---

[1] University Professor, University of El Salvador (UES), El Salvador.
[2] Chief Research Engineer, Department of Structural Engineering, Building Research Institute (BRI), Japan.

the special normative for designing and construction of houses, and another is taken from recorded past earthquakes.

DATA

This research study is based on the experimental study executed to the soil-cement system. Beforehand, sixteen mixtures of soil-cement were widely studied to find out the best mix, taking into account for the selection, parameters like strength, crack pattern, etc. Such proportions of soil-cement are shown in table 1.

Table 1. Soil-cement proportions

Percentage of volume		Proportions
Sand	Silt	Cement-soil ratio
100	0	1:8, 1:10, 1:13, 1:16
75	75	1:8, 1:10, 1:13, 1:16
50	50	1:8, 1:10, 1:13, 1:16
25	75	1:8, 1:10, 1:13, 1:16

Around 70 brick units with dimension of 7x14x28cm were manufactured for each proportion, this means around 1120 units in total. The manufacturing procedure followed was the one proposed by FUNDASAL. The laboratory test carried out for prism and bricks are shown in table 2.

Table 2. Laboratory test of brick unit and prisms

Description	Test	No of specimen
Bricks	Water absorption	80
	Compression	80
	Flexure	80
Prisms	Compression	64
	Diagonal compression	64
Mortar	Compression	36

Besides, four prisms for compression and four for diagonal compression were manufactured for each mix; 128 prisms in total. The mortar used to bond the bricks has a sand-cement ratio of 3:1, which is popularly used to bond bricks. Several scanning procedures were used to select the best soil-cement proportion; even though, the last evaluations were done for four proportions, shown in table 3. Taking into account mechanical properties observed as well as the facility to measure them in field, the proportion using 1:16 cement-soil and 50:50 sand-silt was selected to construct the wall specimens shown and described in table 4.

Table 3. Experimental results in preliminary study

Description	Laboratory Test	Cement:soil ratio (1:13)		Cement:soil ratio (1:16)	
		Average result 25-75	Average result 50-50	Average result 25-75	Average result 50-50
Bricks	Water absorption (%)	28.9	21.8	21.2	34.1
	Compression (MPa)	3.20	7.1	3.60	5.20
	Flexure (MPa)	0.09	0.18	0.07	0.13
Prisms	Compression (MPa)	4.80	4.70	4.80	3.60
	Diagonal compression (MPa)	0.81	0.67	0.81	0.55
Mortar	Compression (MPa)	----	----	22.7	12.7

The reinforcement of the columns and the beams was similar and consisted of 4 longitudinal bars with 9mm diameter, with stirrups of 6mm diameter. The interval of the stirrups was constant in both, beams and columns, and equal to 150mm. The dimensions of concrete sections were 150x150mm for the columns and beams. The panel dimensions were 3x3m approximately and were formed by 36 courses of bricks with 10 units per course. In case of SPP specimen, three walls formed the unit, the lateral walls were similar to the other models and the main panel had beam at the height of 3m.

Table 4. Wall specimens

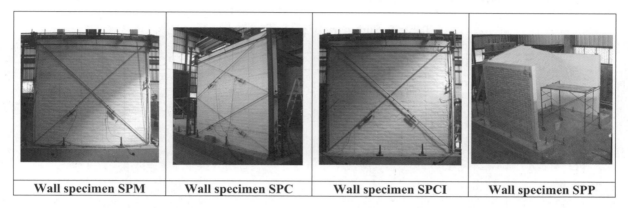

Wall specimen SPM	Wall specimen SPC	Wall specimen SPCI	Wall specimen SPP

Besides, there are soil-cement houses already constructed before and after the current research. The basic model of popular houses has rectangular dimensions of 6.04x4.21m, measured from the exterior edges. It has two doors and two windows as openings in the larger direction. The structural system of the roof consists of channels "C" with 4x2 inches and steel bars with 9mm diameter. The steel bars are welded to the channels to avoid the slipping. Those channels are supported by the ridge beam. In practical construction, the ridge beam is fabricated welding two channels "C". The roof consists of fibrocement sheets placed over the steel bars and channels. The confinement elements with dimensions of 0.15x0.15m and reinforcement as explained above.

THEORY AND METHODOLY

The last decade has witnessed a clear trend towards "performance-based" seismic design, which can be thought of as an explicit design for multiple limit states or performance levels. Analyzing structures for various levels of earthquake intensity and checking some local and/or global criteria for each level has been a popular academic exercise for the last couple of decades, but the crucial development that occurred relatively recently was the recognition of the necessity for such procedures by a number of practicing engineers influential in code drafting. The approach involves the definition of one or more limit states and their associated performance criteria and design checks in a precise and quantitative manner. Seismic limit states recognized in structural design are defined by the damage pattern, which depends on the deformation level reached for structural and nonstructural elements. Therefore, their definition for this system requires the study of parameter like drift, stiffness, resistance and functionality of the structure, which becomes thresholds when the buildings are being evaluated. Three seismic limit states are being proposed in this study: Serviceability, Reparability and Safety. In order to give the best definition for this system, experimental data of in-plane and out-of-plane tests will be analyzed to find out the damage sequence as well as its severity level. Seismic limit states are proposed for each specimen but also proposed for the masonry system. Besides that, two seismic design forces will be applied to the basic house modules with the main targets of checking the seismic design and the prevention of collapse limit state.

RESULTS AND DISCUSSION

The seismic limit states proposed for each specimen has been established through the evaluation of the damage level for all specimens. The results shown in table 5 are similar, SPM and SPC specimens have the same drift level for the serviceability and reparability limit states; however, due to the application of load in only one direction, the collapse drift level is higher for the SPM. Higher drift for each limit state is proposed when the mid-beam is used in construction.

Table 5. Summary of performance levels

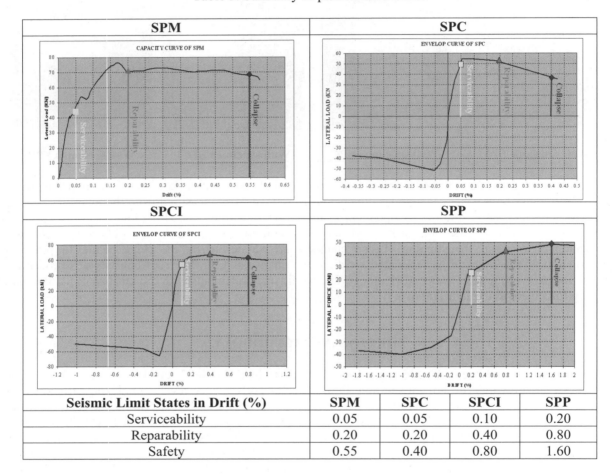

Seismic Limit States in Drift (%)	SPM	SPC	SPCI	SPP
Serviceability	0.05	0.05	0.10	0.20
Reparability	0.20	0.20	0.40	0.80
Safety	0.55	0.40	0.80	1.60

With this element the structure has twice the ductility of SPC, which means that the structure will be less damaged reaching large displacements. In case of the out-of-plane behavior, the drift levels are higher than the drift for in-plane behavior; however, due to differences in definition of both drift, any comparison can be carried out. Even so, it can be said that the lateral failure will occur first, which also agree with the design philosophy where the out-of-plane failure is avoided. After having proposed the seismic limit states, the verification of the collapse house is necessary. To do that, two design seismic forces have been imposed in the in plane and out of plane direction, by using the base shear coefficient provided for the special normative for housing and by using high acceleration observed in past earthquakes. The soil-cement system has shown accepted behavior for both motioned seismic forces in-plane of walls, see table 6. Even by applying the acceleration of 1g, the capacity of the system is enough to withstand the input motions. Damage could be expected for seismic motion with 1g acceleration, but the structure will keep its functionality and the damage will be repaired in short time. In case of the out-of-plane behavior, the serviceability limit state is exceeded by using the recorded acceleration of 1g acceleration. The structure will remain functioning, even when some cracks could have developed in the masonry. Using the base shear coefficient of 0.3 provided for the normative, the structure will behave elastically, within the serviceability limit state.

Table 6. Summary of collapse house prevention

CONCLUSIONS

Regarding the Performance Based Design philosophy, three performance levels or seismic limit states were defined through out this research, which are: Serviceability, Reparability and Safety. Serviceability limit state has been defined as the stage where the structure keeps behaving elastically, minor damage can be expected without harmful effect in the building functionality. In case of Reparability limit state, one particular definition has been chosen for this study, which has to do with the development of steady crack pattern with the first yielding. It is well known that the definition for the reparability limit state several factors have to be evaluated, and that is why its definition range depends on the selected parameters. The Collapse limit state has been defined as the stage where the structure has lost its total capacity, lateral and vertical. However, during the test, this level was not reached due to the existing risk when the wall fall down, then this level has been defined as the stage where the maximum capacity has decreased around 20%, which also involves yielding and severe crack pattern in masonry. The last part of this study was devoted to the verification of the design and prediction of collapse limit state. It can be stated that the house fulfills the requirements established in the normative and also that the house will not reach the collapse for neither of the considered earthquakes. When the house is designed using the seismic demand given by the normative, its in-plane and out-of-plane behavior will keep elastically within the serviceability limit state. However, when strong earthquake struck the structure, both behaviors will keep within the reparability limit state, which means that the collapse will not be reached and also that the damage developed will be economically repairable. The capacity obtained from laboratory test has been higher than both seismic forced, which indicates that the system is strong enough to withstand strong earthquakes.

RECOMMENDATIONS FOR FUTURE RESEARCH

Since the confined masonry system with soil-cement bricks is popularly constructed without technical supervision; booklet shall be prepared to specify simply construction method, which considering reduction factors for designing to take into account the effect of construction quality to keep the house seismic behavior within the margin provided by the different between the capacity/demand ratio observed in laboratory test. During the wall test, differences between the real and predicted capacity were shown up, which have been initially attributed to the workmanship factor as well as the dispersion of quality of materials. Therefore, a properly study shall be carried out to state equations which take into account such variables. Finally, the Performance Based Design philosophy herein explained, shall be applied to the others popular systems as well as the reinforced concrete and masonry building used in El Salvador.

AKNOWLEDGEMENTS

The author wants to express sincere gratitude to Dr Sugano for his support, guidance and friendship during my study. All the investigation presented herein was made possible by the kind cooperation of the Taishin Project staff by providing the necessary experimental data to fully achieve the aims of this study.

REFERENCES

Alvarez -Botero J.C. and López Menjívar M.A., 2004 ,Technical Report
ASIA ,1997, Engineer and architect association of El Salvador.
Astroza I.M. and Schmidt A.A, Revista de ingeniería sísmica No 70 59-75.
ATC-40, 1996, Seismic Evaluation and Retrofit of Concrete Buildings (Volume 1).
CEN, 2003, prEN 1998-1, Eurocode 8.
Crisafulli, F.J. 1997. University of Canterbury, New Zeland.
FEMA273, 1997, NE HRP Guidelines for the Seismic Rehabilitation of Buildings.
FEMA356, 2000, Prestandard and Commentary for Seismic Rehabilitation of Buildings.
FUNDASAL, Salvdorean Foundation for de Development and Minimum Housing, El Salvador.
H. Akiyama, M. Teshigawara, H. Fukuyama, 2000, 12WCEE2000.
Jaramillo, J.D. 2002, Revista de Ingenieria Sismica No. 6753-78. Universidad EAFIT, Colombia.
M. Tomazevic, 1999. Slovenian National Building and Civil Engineering Institute.
Ministry of Public Works (MOP) 1994, El Salvador.
Paulay T and M.J.N Priestly, 1992, United States of America.
SEAOC, 1995, Structural Engineering Association of California, Sacramento, California, US.
T. Paulay and M.J.N. Priestley, 1992. New YorkÑ J. Wiley.
T.P. Tassios, 1984. CIB symposium on Wall Structures, Warsaw.
Taishin Project, 2007. JICA Cooperation project in El Salvador.
Timoshenko S. and Woinowsky-Krieger, 1959, Mc Graw Hill, United States of America.
U.C. Berkeley, 1995. An Action Plan.
Whittier, CA, 1994, 1997, California Office of Emergency Services.

ACCURACY OF SOME GROUND MOTION PREDICTION MODELS FOR PPI STATION-WEST SUMATRA USING OBSERVED STRONG MOTION DATA

Hendarto[*]
MEE07152

Supervisor:Tsuyoshi TAKADA[**]

ABSTRACT

This study proposes the ground motion prediction models applicable to PPI Station located in Padang Panjang, West Sumatra, at which 120 components of strong motion data of 40 earthquake events with sampling rate 100 Hz have been recorded, in order to find site-specific parameters used to investigate the site class and the accuracy of the prediction models for this station. The H/V spectral acceleration ratio scheme is adopted to find the dominant period of ground beneath this station since there are no quantitative subsurface soil properties available. The results of H/V analysis show that the mean peak of H/V ratio of EW, NS and total horizontal components to vertical component is around 0.2 second. The peak period of H/V ratio is in good agreement with the geologic information from geological surface map published by Geological Research and Development Center (GRDC, 1973). The geologic formation for the ground beneath and around this station is dominated by quaternary volcanic rock (Qast). Both results indicating the site class for the ground beneath the station could be categorized as rock site with AVS30 more than 600 m/s. The existing prediction models for peak ground acceleration (PGA) and 5% damped spectral acceleration (Sa) at rock site are tested in this study. The accuracy of the selected models is discussed on the basis of the statistical distribution of the logarithmic deviation between the prediction and the observed value. The mean residuals of all prediction models of Sa are found to have the same tendency that they underestimate Sa at period T = 0.2 second and overestimate Sa at T $\geq$ 0.4 second. It is found that all prediction models show significant period-dependent mean errors and are not allowed to be applied to this site without site correction factors, and among them, Youngs et al., 1997, Kanno et al., 2006, and Zhao et al., 2006 provide the smallest prediction error. Based on these results, it can be concluded that all prediction models may be applied with each site correction factor to reflect the site-specific condition and among them three models proposed by Youngs et al., 1997, Kanno et al., 2006, and Zhao et al., 2006a are the best for this station. However, the error is still large; therefore a further study is needed to check it.

Keywords: peak ground acceleration, 5% damped spectral acceleration, mean residuals.

INTRODUCTION

One of the important factors in the seismic hazard assessment for certain regions is the selection of ground motion prediction model which is also known as attenuation relation model.

There are many attenuation relation model determined by the different definitions of independent variables such as distance, magnitude, site condition, types of faulting and selection of the horizontal ground motion component. Unfortunately, there is no available attenuation model derived

---

[*]State Ministry of Research and Technology, Indonesia.
[**] Professor, Graduate School of Engineering, the University of Tokyo, Japan.

from using strong motion databases from Indonesia despite Indonesia is located at active tectonic region.

We adopt only some attenuation relation models in both probabilistic and deterministic seismic hazard calculation. However, the selection of suitable attenuation relation for a certain area often causes serious practical problem and doubt since there are more than 206 attenuation relation models to predict PGA and 127 attenuation models to predict response spectra ordinate (Douglas, 2006).

This study will emphasize the scheme in selecting suitable attenuation relation models from some model candidates and also find the site factor for a particular site to account the local site characteristic of the site based on the selected attenuation relation model. Site factor is the mean value of the logarithmic residuals between the observed and predicted values, and evaluated it with the limited observations (Morikawa et al., 2006).

DATA

120 components of qualitative records with sampling rate 100 Hz from 40 events at PPI station used in this study are provided by JISNET-NIED with permission from Meteorological and Geophysical Agency of Republic of Indonesia (BMG Indonesia).

The basic assumptions of earthquake source parameters such as epicenter location, focal depth, magnitude and focal mechanism are referred to CMT and PDE-USGS catalog. Most events, 33 of 40 events are referred to CMT catalog, while the information of other 7 events is referred to PDE-USGS catalog.

39 of 40 events were triggered by subduction earthquake sources. Another one event was triggered by shallow crustal source.

THEORY AND METHODOLOGY

Magnitude Scale

It is an important consideration to use a unified magnitude scale in attenuation study. Therefore in this study, the relation of mb, M_S and M_W is adopted from Scordilis, 2006 in order to unite magnitude scale. By applying these equations, the homogeneity of database will be provided.
Relation between M_W and M_S, for $3.0 \leq M_S \leq 6.1$

$$M_W = 0.67(\pm 0.005)M_S + 2.07(\pm 0.03)$$ (1)

Relation between M_W and m_b, for $3.5 \leq m_b \leq 6.2$

$$M_W = 0.85(\pm 0.04)m_b + 1.03(\pm 0.23)$$ (2)

Attenuation Relation Models

We select five models for spectral acceleration (Sa) since we cannot cover all existing models in this study. The five models for predicting Sa are Atkinson and Boore, 1997 (AB97); Kanno et al., 2006 (K06-S and K06-D); Midorikawa and Uchiyama, 2006 (MU06-S and MU06-D); Youngs et al., 1997 (Y97-Inter and Y97-Intra); and Zhao et al., 2006a (Z06).

The selected period range for this study is from PGA to $T \leq 2$ second based on the most building types, geometries and structures constructed around the particular site (i.e. PPI seismic station) which are dominated by concrete structure buildings with the total number of stories vary from 1 to 8-story buildings.

Site Classification Method

There are some previous studies conducted by several researches related to the determination of site classification with different methods. Two parameters that commonly used in those studies are the average shear wave velocity of top 30 m (AVS30, V_{30} or $\overline{V}_S$) and the dominant period of site (T_G). The relation between them is described in many codes such as NEHRP and Japan Road Association, 1980.

Zhao et al., 2006b used V_{30} (using 0.25 times the site period in four site classes as the shear travel time in the top 30 m soil layers, see Eq. (3)) and site classes used in the Japan Road Association, 1980 and NEHRP Provision (see Table 1).

In this study, H/V response spectral ratio scheme proposed by Zhao et al., 2006 will be adopted to find the dominant period (T_G) of site of PPI seismic station located. The result of H/V scheme will be cross-checked by surface geology around PPI Station based on geology maps of Padang provided by Geological Research and Development Center (GRDC, 1973), Republic of Indonesia.

$$V_{30} = \frac{30}{0.25.T_G} \qquad (3)$$

Table 1. Site class definitions used in Japan for engineering design practice and the approximately corresponding NEHRP site classes (Zhao et al, 2006b)

Site class		Site natural period (sec)	Average V_{30} (m/s)	NEHRP class
SC I	Rock/stiff soil	$T_G < 0.2$	$V_{30} > 600$	A+B
SC II	Hard soil	$0.2 \leq T_G < 0.4$	$300 < V_{30} \leq 600$	C
SC III	Medium soil	$0.4 \leq T_G < 0.6$	$200 < V_{30} \leq 300$	D
SC IV	Soft soil	$T_G \geq 0.6$	$V_{30} \leq 200$	E

RESULTS AND DISCUSSION

Site Class

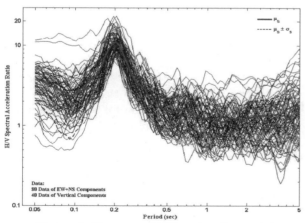

Figure 1. The mean H/V ratio of spectral acceleration at PPI station.

Based on the H/V analysis result, the mean peak ratio of EW and NS components to vertical components is around 0.195 and 0.205, respectively. Figure 1 shows the mean peak ratio of total components to vertical components. By referring to Table 1, the site class beneath PPI Station could be SC I (class A or B) or SCII (class C).

As mentioned earlier, another way as basic engineering judgment to determine site class is that by using the surface geological condition of site beneath and around PPI Station.

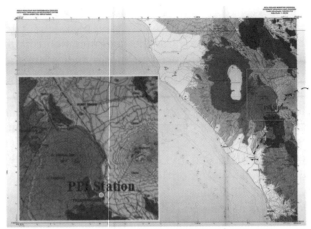

Figure 2 shows that the structural geology beneath and around this station is dominated by volcanic rocks (andesit rock, Qast). By referring to NEHRP classification, we may make a preliminary estimate that the site class beneath this station is probably class C (very dense/soft rock) or B (rock). Based on the H/V spectral acceleration ratio and the geological surface condition, the site classification of PPI station can be assumed as rock site (SC I or with the average shear wave velocity of top 30m is predicted more than 600 m/s.

Figure 2. Geology map around PPI station (GRDC, 1973).

Selecting Candidates

The critical question arisen is how to select appropriate attenuation model for a particular site, i.e. PPI Station. We did statistical analysis for answering that question based on the mean residuals. There are several different ways that can be used to construe the results of the statistical analysis. The first method is to select the best model by using both the lowest mean residual (see Figure 3) and standard deviation (Figure 4) at the required and specific period. The second method is to select two or three satisfactory models and average the predicted ground motions. The last method is to adjust all prediction models by applying a correction factor based on the mean residuals.

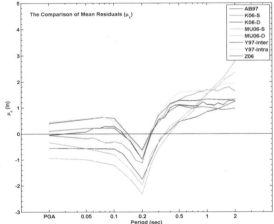

Figure 3. Mean residuals of Sa for all attenuation relation models.

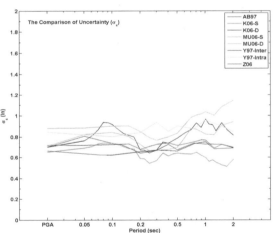

Figure 4. Uncertainty of Sa for all attenuation relation models.

Generally, the decision in selecting the best model will be depended on the application of interest. Related to the first method, some findings can be used as guidance to justify which one is the most appropriate among the used models in this study. The important thing based on the mean residuals (see Figure 3) is that all models have the same tendency which underestimated to predict Sa at period $T = 0.2$ second and overestimated to predict Sa at $T \geq 0.4$ second. At the short period range, Zhao et al., 2006 and Kanno et al. 2006 attenuation relation models provided smaller mean residuals for PGA and $T < 0.1$ second (relatively), than the other models did; and Youngs et al., 1997 for both interface and intraslab events obtained the smallest mean residuals among others models for period $0.15 \leq T < 0.25$.

Another important finding is that the comparison of uncertainties presented in Figure 4 for various period ranges. It can be concluded that at the short period, for PGA and T < 0.2, Youngs et al,. 1997 attenuation relation model for interface events has the smallest uncertainty value while Kanno et al., 2006 attenuation relation model for shallow events has the smallest one for period T ≥ 0.2 second. We may say that Youngs et al., 1997, Kanno et al., 2006 and Zhao et al., 2006 are suitable for this station. Note that 39 of 40 events used in this analysis were triggered by subduction sources, hence we could not say too much about the prediction of ground motions for crustal event since only one event is available in the data base.

Site Correction Factor Based on Mean Residuals

The site correction factors can be defined as the factors determined by the average residuals treating the adjusted attenuation relation model as a site-specific model. The relation between the original and the adjusted model can be expressed by Eq. (4).

$$\ln y(t)^* = \ln y(t) + s(t)$$

$$\text{or} \tag{4}$$

$$\log y(t)^* = \log y(t) + s(t)$$

Where $y(t)^*$ is the adjusted model, $y(t)$ is the original model and $s(t)$ is the site correction factor at period t. The site correction factors for all prediction models are represented in Figure 5.

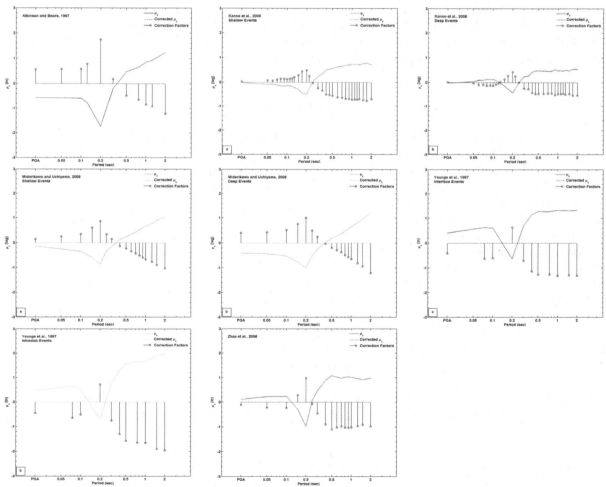

Figure 5. Site correction factors for various attenuation relation models.

CONCLUSIONS

Site correction factors for all models are determined based on the value of mean residuals. Note that these corrections are made only for mean predictions, not for uncertainty predictions. We can adjust the predicted value from each model for PPI station by adding these correction factors into the base models of each attenuation relation model, (see Eq. (4)). Then all prediction models may be applied with each site correction factor to reflect the site-specific condition and among them three models proposed by Youngs et al., 1997, Kanno et al., 2006, and Zhao et al., 2006a are the best for this station. However, the error is still large; therefore a further study is needed to check it.

RECOMMENDATION

As future studies, estimation of site correction factors for other sites, development of possible methods for reducing prediction errors which are still large, and nation-wide deployment procedure to develop site-specific ground motion prediction models.

AKNOWLEDGEMENT

I would foremost like to thank Dr. Yoshimitsu Fukushima for our scientific relationship in the preparation and data treatment; Dr. Fauzi from BMG Indonesia, Dr. Hiroshi Inoue from JISNET Network and Dr. Winfried Hanka from GFZ-Potsdam for strong motion data collection. I also thank Dr. Anthony Lomax and Dr. Toshihide Kashima for either direct or indirect discussion related to SEISGRAM2k and Viewave; Wessel, P. and Smith, W. H. F. for the GMT software; and my advisor, Dr. Shin Koyama, for giving his time in our fruitful discussion

REFERENCES

Atkinson, G. M., and Boore, D. M., 1997, Seismological Research Letters, 68, 74–85.

Douglas, J., 2006, BRGM/RP-56187-FR.

Fukushima, Y., and Tanaka, T., 1990, Bulletin of the Seismological Society of America, 80, 757-783.

Geological Research and Development Center (GRDC), 1973, Quadrangle Padang, 0715

Japan Road Association, 1980, Maruzen Co., LTD.

Kanno, T., Narita, A., Morikawa, N., Fujiwara, F., and Fukushima, Y., 2006, Bulletin of the Seismological Society of America, Vol. 96, No. 3, pp. 879–897.

Midorikawa, S., and Ohtake, Y., 2004, 13th World Conference on Earthquake Engineering Vancouver, B.C., Canada August 1-6, Paper No. 325.

Morikawa, N., Kanno, T., Narita, A., Fujiwara, H., and Fukushima, Y., 2006, Japan Association for Earthquake Engineering, vol. 6, No. 1 (in Japanese).

NEHRP, 2000, FEMA 368/369, Washington, D. C.

Scordilis, E. M., 2006, Journal of Seismology, 10, 225-236.

Uchiyama, Y. and Midorikawa, S., 2006, Journal of Struc. and Const. Eng., AIJ, 606, 81-88.

Youngs, R. R., Chiou, S. J., Silva, W. J., and Humphrey, J. R., (1997), Seismological Research Letters, 68, 58-73.

Zhao, J. X., Zhang, J., Asano, A., Ohno, Y., Oouchi, T., Takahashi, T., Ogawa, H., Irikura, K., Thio, H. K., Somerville, P. G., Fukushima, Ya., and Fukushima, Yo., 2006a, Bulletin of the Seismological Society of America, Vol. 96, No. 3, pp. 898–913.

Zhao, J. X., Irikura, K., Zhang, J., Fukushima, Y., Somerville, P. G., Asano, A., Ohno, Y., Oouchi, T., Takahashi, T., and Ogawa, H., 2006b, Bulletin of the Seismological Society of America, Vol. 96, No. 3, pp. 914–925.

EVALUATION OF SHEAR STRENGTH OF PRESTRESSED CONCRETE BEAMS WITH HIGH STRENGTH MATERIALS

Ferri Eka Putra[*]   **Supervisor: Susumu KONO**[**]
MEE07159

ABSTRACT

Shear behavior of prestressed concrete beams has been studied for many years and some equations were already proposed in several codes. However, these equations do not correctly evaluate which resulting in conservative design or dangerous design. One hundred sixty-eight (168) data from 14 papers published in Japan during 1981 to 2004 are used in analysis. Equations provided by Japanese PC standard 1998 and ACI 318-05 are studied using these data. Based on analysis, equation 71.2a of Japan PC standard 1998 was the best design equation to predict shear strength of prestressed concrete beams among six equations. However, the equation has following shortcomings. It is quite conservative for I-section and un-bonded rectangular section. On the other hand, it overestimate shear strength of pre-cast prestressed concrete and rectangular beam using high strength steel as shear reinforcement. To improve its performance, the arch contribution was isolated by choosing beams with un-bonded tendon or specimens without shear reinforcement. Then, the modification was made to; the effect of prestressing force on the compressing strength of concrete strut, confining effect provided by shear reinforcement on the compressive strength of concrete strut and the effective width of the web for I-sections. On the other hand, truss contribution was not modified except for I-section. Proposed modification has produced satisfying results. Average experiment-to-prediction ratio is getting closer to 1.0 and scatter of data is getting small.

Keywords: Prestressed concrete, Japanese PC standard 1998.

INTRODUCTION

Use of prestressed concrete (PC) has became popular for building structure recently because it provides advantages such as high performance, high durability and other benefits from economic point of view. In developed country such as America and Japan, this type of structure already applied for many types of buildings such as residents, offices, universities, hospitals and sports facilities.

On the other hand, research in PC field has been conducted by several institutions and researchers around the world. Moreover, some equations were already proposed to predict behavior of PC beams. However, these equations do not correctly evaluate especially for shear behavior. Comparing with flexural behavior which is already able to be predicted easily, the exact shear behavior of member is still unpredictable. Due to that reason, most of codes in the world prefer to use higher safety factor due to accommodate the uncertainty behavior of shear. Therefore, evaluation of equation provided by code should be done to achieve more efficient design.

The purpose of this thesis is to gain more understanding about shear behavior of PC beams and to propose some consideration for modifying existing standard to get more efficient and reasonable result in predicting shear strength of PC beams

---

[*]Research and Development Center for Human Settlements, Department of Public Works, Indonesia
[**]Assoc. Professor, Graduate School of Engineering, Kyoto University, Japan.

STATE OF THE ART

Shear Mechanism

According to the discussion of America Concrete Institute (ACI) committee 445 in 1998, shear resistance capacity for concrete members without shear reinforcements is contributed by five components, namely un-cracked concrete and flexural compression zone (V_c), aggregate interlock (V_a), dowel action of longitudinal reinforcement (V_d), residual tensile stress across crack and Arch action. The amount of contribution will depend on geometry of members, type of loading, steel ratio, type of web reinforcement and prestressed reinforcement (CTA, 1976). On the other hand, shear reinforcement plays important role to ensure that shear failure will not occur. The contribution of shear reinforcement is to improve contribution of dowel action, suppressing flexural tensile stresses, limiting the opening of diagonal cracks, providing confinement and preventing the breakdown of bond when splitting cracks develop in anchorage zones (Park and Paulay, 1997).

Equation 71.2a of Japanese PC Standard 1998

Equation 71.2a (presented by eq.1) was developed based on superposition of arch and truss mechanism. Left part of equation presents truss mechanism and right part presents arch mechanism.

$$Q_u = b_0 j_0 p_{ww} f_y + \frac{b_0 D}{2} \left(v f'_c - 2 p_{ww} f_y \right) \tan \theta \qquad (1)$$

where, b_0 is width of beam, j_0 is distance between compressive and tensile strength, D is depth of beam. Besides, p_w is reinforcement ratio of stirrup, f'_c is compression strength of concrete, v factor is effectiveness factor. This factor is influenced by the shear span to depth ratio, axial force which represented by effective prestressed stress and compressive strength of concrete. It amount is limited to the range of 0.65 to 1. This limitation introduced to represent maximum, 100 % of compressive strength of concrete can contributes in arch mechanism. Besides, shear reinforcement effect is considered in terms of shear reinforcement ratio.

DATABASE

There are 14 prestressed concrete research had been done during 1981 to 2004 in Japan and at least 168 specimens were tested. 50 specimens are prestressed beam without shear reinforcement and 118 specimens with shear reinforcement. Compressive strength of concrete which was used for the specimens is varied from 27 MPa to 110 MPa. The ratio of longitudinal reinforcement varies from 0.35 to 2 %. The ratio of prestressed reinforcement is varying from 0.2 up to 3.6 % and ratio of shear reinforcement is from 0.2 to 1.4 %. The shear spans to depth ratio of specimen varies from 1 to 3.2. Besides, specimens can be divided into two group; rectangular shape and "I" shape. At least 107 specimens are rectangular shape section and 60 specimens are "I" shape section. Rectangular section can be divided into 2 types; bonded prestressed reinforcement (90 data) and un-bonded prestressed reinforcement (17 data). Meanwhile, "I" shape section only use bonded prestressed reinforcement. Data in this study were compiled from journal as shown in table 1.

Table 1. Source of database

Author	Publisher	Volume and Page	Year
Funakoshi M, Okamoto T	1	Vol.3, 365-368	1981
Funakoshi M, Okamoto T, Tanaka E	1	Vol.4, 297-300	1982
Fukui, Ookuma A, Hamahara M, Suetsugu H,	2	C-2, Structures IV, 877-880	1996
Fukui T, Nagasawa T, Hamahara M, Suetsugu H	2	Structures II, 1023-1028	1994
Funakoshi M, Tanaka H, Tani M	1	Vol.6, 469-472	1984
Matsizaki Y, Hirayama A, Kobayashi A, Sakai M	2	Vol.59, 1683-1684	1984

Table 1. Source of database (continued)

Author	Publisher	Volume and Page	Year
Matsizaki Y, Suzuki K, Matsutani T, Shirahama S, Takayama Y	2	C-2, Structures IV, 887-890	1996
Muguruma Hiroshi, Watanabe Fumio, Fujii Masanori	2	Vol.58, 2549-2550	1983
Ogawa T, Saito A, Iida S, Fukui T, Suetsugu H, Sakiyama K, Hamahara M	2	C-2, Structures IV, 1077-1082	1999
Ohtaka K, Yuasa N, Hamahara M	2	C-2, Structures IV, 991-994	2002
Ookuma A, Fukui T, Hamahara M, Suetsugu H	2	C-2, Structures IV, 847-852	1997
Wakamatsu S, Takizawa K, Takagi H, Shiraishi I	2	C-2, Structures IV, 1039-1044	1998
Yuasa N, Ogawa T, Kamakura M, Fukui T, Uchida R, Hamahara M		C-2, Structures IV, 965-970	2000
Yuasa N, Ohtaka K, Fukui T, Hamahara M	2	C-2, Structures IV, 955-960	2001

[1] Proceedings of the Japan Concrete Institute.

[2] Summaries of technical papers of Annual Meeting Architectural Institute of Japan.

RESULTS AND DISCUSSION

Modification of the Arch Contribution in Equation 71.2a of PC standard 1998

In general, equation 71.2a of PC standard 1998 shows best performance in predicting behavior of shear strength among six studied equations. However, the equation has following shortcomings and some modification should be done for applying this equation to rectangular un-bonded and "I" shape section beams to get more efficient design.

Modification of the arch contribution for rectangular un-bonded beams

Truss mechanism can not exist for un-bonded specimen because it comes from internal force which resulted from bonded condition. Therefore, absence of bond will distinguish truss mechanism. Figure 1 show specimens without shear reinforcement or with $p_w = 0\%$ has ratio of experiment-to- prediction in range of 1 to 1.75. Average experiment-to-prediction ratio is 1.5 and it standard deviation is 0.5. Therefore, some modification should be done to get more accurate prediction result. Circled data in figure 1 will be neglected for next analysis since it has different specification than other data.

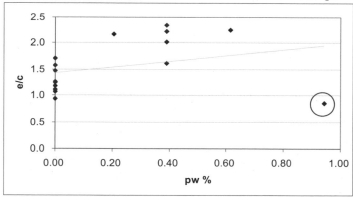

Figure 1. Performance of equation 7 with varying shear reinforcement ratio for un-bonded rectangular section specimens

The effect of axial force of prestressed reinforcement could be one factor which produces higher contribution of compressive strength of concrete in arch mechanism. Therefore, term of ν factor in eq. 71.2a to represent the contribution of axial force may not be correctly evaluated. For instant, neglect the upper limitation of ν factor as shown in eq. 2 can be used to represent higher contribution of axial force. Meanwhile, bottom limit of this equation is kept to guarantee at least 65 % of its compressive strength will contribute in resisting shear force. Proposed equation to obtain modified ν factor is presented by eq. 2.

$$ v_{\text{mod}} = \alpha\, L_r \left(1 + \frac{\sigma'_g}{f_c} \right) \quad 0.65 \le v_{\text{mod}} \tag{2} $$

Contribution of shear reinforcement in resisting shear for un-bonded specimen could be neglected due to the absence of bond which play important role for truss mechanism. However, based on experiment result, the contribution of shear reinforcement still exists. It is expected that shear reinforcement confines the core of concrete and compression strength of the concrete increases Amount of shear reinforcement contribution in un-bonded specimens can be solved by using eq. 3. However, advanced research should be done to evaluate this equation.

$$c_1 = 1.7\ pw + 1 \qquad (3)$$

where, c_1 is coefficient of confining contribution and p_w is shear reinforcement ratio in percent.

Finally, we propose equation 4 as modification of equation 71.2a to predict more accurate of shear strength rectangular un-bonded beams.

$$Q_u = \frac{b_0 D}{2} c_1 \nu_{mod}\ f'_c \tan \theta \qquad (4)$$

By using equation 4, prediction result becomes closer to experiment result. Average ratio of experiment-to-prediction decrease from 1.5 to 1.1 and it standard deviation decrease from 0.50 to 0.13. It can be said that our proposal is working well to predict shear strength of un-bonded PC beams. Figure 2 shows performance of proposed equation for un-bonded rectangular section.

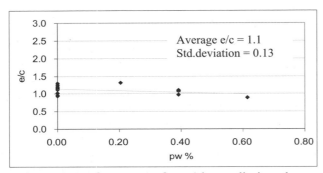

Figure 2. Performance of eq. 4 in predicting shear strength of un-bonded rectangular section specimens

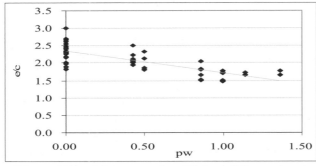

Figure 3. Performance of eq.71.2.a with varying shear reinforcement ratio for I section specimens

Modification of the arch contribution for "I" shape section beams

Based on figure 3, we can observe that the accuracy of this equation is getting worse for $p_w = 0$ %. Ratios between experiment and prediction are located in range 2 to 3. It means the results are too conservative. For specimens without shear reinforcement, shear strength of PC member is only contributed by arch action.

Therefore, we can assume that equation for arch action part is not correctly evaluated. Based on statistic approach, we get an average of experiment-to-prediction ratio is 2.1 and standard deviation is 0.34. We were trying to apply the same modification for ν factor as well as we did for un-bonded specimens. Equation 2 is used to replace original term of ν factor in original equation. On the other hand, we are modifying term of b_0 since using this term in calculation produced conservative results.

We found that ratio of gross section area to the depth of specimens is resulting best result to replace term of b_0. This ratio can be called as effective width (b_{eff}). Finally, we proposed equation 5 to predict shear strength of I-section PC beams.

$$Q_u = b_{eff}\, j_0\, p_{ww}\, f_y + \frac{b_{eff}\, D}{2}\left(v_{mod}\, f'_c - 2\, p_{ww}\, f_y\right)\tan\theta \qquad (5)$$

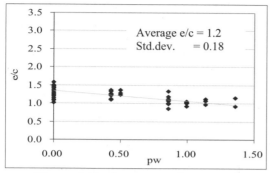

Figure 4. Performance of proposed modification (eq.5) for "I" section specimens

Equation 5 produces better result. Average experiment to prediction ratio was reduced from 2.1 to 1.2 and it standard deviation was reduced from 0.34 to 0.18. Moreover, data of experiment to prediction ratio which less than 0.95 is less than 5%. Therefore, this proposal is efficient and safe to be adopted. Figure 4 shows performance of equation 5. We can observe that the results are more consistent with specimens with and without shear reinforcement which represented by small gradient of its results. This tendency shows arch and truss mechanisms are represented well by using this consideration.

Rectangular bonded beams

Equation 71.2a works well in predicting shear strength of bonded rectangular specimens. Average ratio of experiment-to-prediction result is 1.0 and its standard deviation is 0.22. However, 16 of 90 specimens can not be predicted well by this equation. The equation tends to overestimates shear strength of specimens which using high strength steel (960 MPa). On the other hand, from 8 data which can be collected, all of them have the ratio of experiment-to-prediction results in range of 0.5 to 0.6. Another situation which could not be predicted well by equation 71.2a is using of pre-cast system. Equation 71.2a tends to overestimate shear strength of pre-cast specimens using strand 70 as prestressed tendon. Three of 4 specimens have shows this tendency. The ratio of experiment to prediction result for these specimens is around 0.75 to 0.85. Besides, for pre-cast system using high strength bar as prestressed reinforcement, equation 71.2a produces conservative results.

Based on previous approach in increasing performance of equation 71.2a, similar modification was applied to bonded rectangular section specimens. Term of modified "v" factor which obtained from equation 2 was used. Therefore, modified equation 71.2a can be written as eq. 6.

$$Q_u = b_o\, j_0\, p_{ww}\, f_y + \frac{b_o\, D}{2}\left(v_{mod}\, f'_c - 2\, p_{ww}\, f_y\right)\tan\theta \qquad (6)$$

Figure 5 show performance of equation 6 for monolithic rectangular bonded specimen only.

Equation 6 produce slightly change in predicting shear strength of rectangular bonded specimen. By using equation 6, average experiment-to-prediction ratio was reduced from 1.08 to 1.00 and it standard deviation is 0.14. Therefore, equation 6 is proposed only for monolithic rectangular bonded PC beams. Figure 5 shows performance equation 6 with varying shear reinforcement ratio for monolithic bonded rectangular beams.

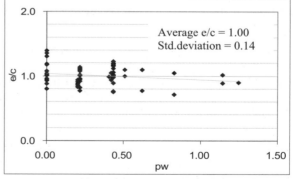

Figure 5. Performance of equation 6 for monolithic bonded rectangular section specimens

CONCLUSIONS

Shear behavior of prestressed concrete beams has been studied for many years and some equations were already proposed by several codes. However, these equations were not correctly evaluated. One hundred sixty-eight (168) data from 14 papers published in Japan during 1981 to 2004 were used in analysis. The equations provided by Japanese PC standard 1998 and ACI 318-05 were studied using these data. Based on analysis, equation 71.2a of Japan PC standard 1998 was the best design equation to predict shear strength of prestressed concrete beams among six equations. Average of experiment-to-prediction ratio for all types of specimens is 1.8. Besides, its standard deviation is 0.6. However, the equation has following shortcomings. It is quite conservative for I-section and un-bonded rectangular section. On the other hand, it overestimates shear strength of pre-cast PC and beam using high strength shear reinforcement.

In order to improve the performance of equation 71.2a, the arch contribution was isolated by choosing beams with un-bonded tendon or specimens without shear reinforcement. The modification was made to; the effect of prestressing force on the compressing strength of concrete strut, confining effect provided by shear reinforcement on the compressive strength of concrete strut and the effective width of the web for I-sections. Meanwhile, truss contribution was not modified except for I-section. Equations for I-sections and rectangular section were separated and then modifications for each type of beam were proposed. For rectangular un-bonded specimens, we proposed neglecting of upper limit of v factor and considering confining effect by shear reinforcement. Average ratio of experiment-to prediction for proposed modification has decreased from 1.5 to 1.1 and it standard deviation has decreased from 0.50 to 0.13. For I-shape section, using of effective width and neglecting upper limit of v factor were proposed. Effective width can be obtained from ratio of gross area of section by depth of beams. Its average experiment-to-prediction ratio was reduced from 2.1 to 1.2 and it standard deviation was reduced from 0.34 to 0.18. For rectangular-bonded specimens, equation 6 is proposed. However, the equation is valid only for monolithic rectangular bonded beams. Average experiment-to-prediction ratio was reduced slightly from 1.08 to 1.00 and it standard deviation is remain 0.14

RECOMMENDATION

Some modifications have already been proposed to increase performance of equation 71.2a of Japan PC Standard 1998. However, some conditions still can not be satisfied by proposed modifications. Shear strength of specimens using of high strength shear reinforcement and shear strength of pre-cast PC beams needs to be evaluated. Moreover, contribution of axial force provided by prestressing force is not clearly determined. Furthermore, truss contribution was not evaluates except for I-section. Therefore, further study to observe these conditions should be done.

AKNOWLEDGEMENT

I would like to express my heartiest gratitude to my advisor, Dr. Shin Koyama for giving very useful advices during my individual study. I also would like to thanks to my colleague in Kono's Laboratory of Kyoto University, Ms. Y Ichioka and Dr. M Sakashita who help me learn Japanese code.

REFERENCES

American Concrete Institute, 2005, ACI.
American Concrete Institute, 1998, ACI
Architectural Institute of Japan, 1994, AIJ
Architectural Institute of Japan, 1998 AIJ
Concrete Technology Associates, 1976, Technical Bulletin 76-B11/12
Park, R. and Paulay, T, 1973, John Wiley & Sons

SEISMIC DESIGN OF THE MASONRY INFILL RC FRAME BUILDINGS WITH FIRST SOFT STOREY

Prakirna TULADHAR *   Supervisor: Dr. Koichi KUSUNOKI**

MEE07158

ABSTRACT

The main aim of this study was to investigate the seismic performance and design of the masonry infill Reinforced Concrete (R/C) frame buildings with the soft first storey under a strong ground motion. The study also highlighted the error involved in modeling of the infill RC frame building as completely bare frame neglecting stiffness and strength of the masonry infill wall in the upper floors. The envelope curve for representing non-linear behaviors of masonry infill wall was determined for numerical analysis. The attempt was made to determine the strength increasing factor (η) to account the effect of the soft storey through various 2D analytical models using Capacity Spectrum Method (CSM) and established its relationship with initial stiffness ratio (K_o2/K_o1). In this study, the non-linear dynamic time history analysis was also carried out with a three dimensional practical model in order to verify the proposed strength increasing factor (η).

Key words: First soft storey, Capacity Spectrum Method, Strength increasing factor, Storey drifts

INTRODUCTION

In urban and semi-urban areas of Nepal, like several other developing countries around the world, reinforced concrete frame structure is very popular for the building construction. Usually, the RC frame is filled with burnt clay bricks masonry as non-structural wall for partition of rooms in the floors which is called infill wall. The reason behind the popularity of using brick masonry in developing countries may be due to economy, architecture; aesthetic, locally available skill and material, good for water proofing, heat and sound insulation, and better security. Similarly, open first storey is nowadays as unavoidable features for the most of urban multistory buildings for vehicles parking, shop, reception, and lobby etc. In building with the first soft story, the upper storey being stiff undergoes less inter-storey drift, while the displacement will be concentrated at the first storey during seismic events. Hence, an abrupt change in the stiffness in first storey have adverse effect on the performance of the building during ground shaking which is known as first soft storey effect.

Problem statement

In current practice of structural design of the buildings in Nepal, masonry infills are treated as non-structural element and the frame is modeled as a bare frame. The strength and stiffness contribution of infill walls are often neglected and self weight of infill wall is modeled as uniformly distributed load. The entire lateral load is assumed to be resisted only by the moment resisting frame. In reality, the presence of the infill wall in the frame changes the lateral load transfer mechanism into truss actions leading to stress concentration in the ground floor. Hence, the effect of the first soft storey is failed to capture in the analysis. This error in modeling may lead to damage in the first soft storey during a strong earthquake.

---

* Engineer, Dept. of Urban Development & Building Construction, Govt. of Nepal
** Associate Prof., Yokohama National University, Yokohama, Japan

Macro modeling for brick infill masonry wall

The contribution of the masonry infill wall to represent the infill frame can be modeled by replacing the panel by a system of equivalent horizontal shear spring connected to the main frame vertically. The non-linear properties of the equivalent shear spring can be determined by the envelope curve. It can be represented by poly-linear strength envelop with initial elastic stiffness until the cracking force Q_c , post cracking degraded stiffness until the maximum force Q_y and post peak residual force Q_u. The corresponding lateral displacements are γ_c, γ_y and γ_υ respectively (Ref. Figure 1).

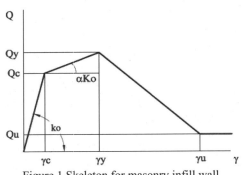

Figure 1 Skeleton for masonry infill wall context.

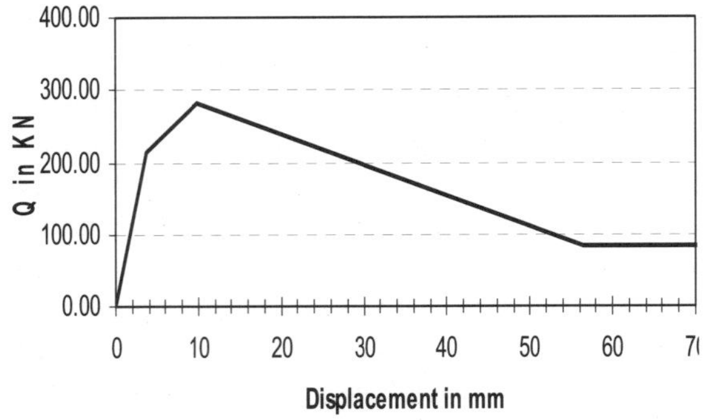

Figure 2 Skeleton curve for masonry wall in Nepalese

In this study, the brick masonry wall in Nepalese context was considered to determine the skeleton curves with the given parameters: Size of brick = 230x110x57mm, thickness of mortar= 10mm and mortar ratio 1:5. The compressive strength of the burnt brick, f_{cb}= 7.5 N/mm^2, compressive strength of mortar , f_j= 5.0 N/mm^2 and tensile strength of brick, f_{tb} = 0.1f_{cb} = 0.75 N/mm^2. The computer program was written in FORTRAN to obtain strength envelope (Skeleton) curve as shown in figure 2.

NUMERICAL MODLING FOR THE STRUCTURAL ELEMENTS

The structural elements were modeled by STERA3D (Dr. Taiki Saito, the chief researcher of IISEE, BRI) for non-linear structural analysis. All the structural members were considered as line elements with non linear springs to represent the non-linear behavior of the structure. The cross sections were considered to be rigid. The details of modeling are discussed below.

Column: The structure to be analyzed being first soft storey and often subjected to large fluctuating axial load due to overturning moment, the fiber model (multi spring model) was used for the column element to take into account the interaction between bending moment and axial force. Column element was idealized with the line element with fiber slice model at each ends consisting of steel and concrete multi axial springs elements in order to represent flexural and axial deformation to take account of the interaction between bi-directional bending moments and axial load. The bi- axial shear springs were also considered to represent the non linear shear deformation of the column. An origin-oriented poly-linear hysteresis model was used for representing non-linear shear spring.

Beam: The beam elements were represented as a uni-axial model with non-linear rotational spring at two critical ends and non-linear shear spring at the middle of the beam for shear deformation. The hysteresis tri-linear degradation modified Takeda model was applied for non-linear rotational spring model

Masonry infill wall: Masonry walls were represented with a line element with non-linear shear spring at the mid of element. The poly-linear slip model was used as the hysteresis model.

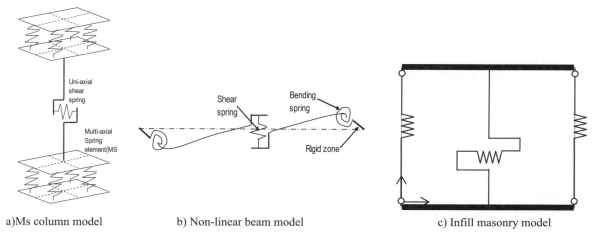

| a)Ms column model | b) Non-linear beam model | c) Infill masonry model |

Figure 3 Structural Elements Models

DETERMINATION OF STRENGTH INCREASING FACTOR (η) FOR FIRST SOFT STOREY BY CAPACITY SPECTRUM METHOD

It was already discussed about soft storey effect in first storey due to presence of infill wall in the upper floors. However, the designers do not consider the strength and stiffness of the masonry wall in their analysis due to limitation in their knowledge. The performing the non-linear analysis is also very difficult for many engineers. Hence, the attempt was made to determine the strength increasing factor (η) to account for the soft storey effect due to infill wall and tried to establish its relationship with initial stiffness ratio between two floors and height of the structure so that designer can increase the lateral strength of the columns subjected to first soft storey to meet the drift demand due to the soft storey.

Model studied:

A prototype apartment building with typical floor plan was considered. The building was symmetrical in plan and the plan dimension of the building was 20m x 12m. The span length along both the axis was 4m c/c and floor height 3m c/c. The structural frame elements were designed for six storeys according to current practice with critical load combination prescribed in the prevailing code.

Capacity Spectrum Method:

Capacity Spectrum Method (referred as to CSM hereafter) can provide the realistic representation of the structure subjected to a severe earthquake and can be established as a rational alternative to nonlinear analysis for assessing seismic demands of the MDOF structure. The seismic demand in form of response spectrum and capacity in form of pushover analysis is plotted in ADRS format. The point of the intersection of two curves is the response of structure for the demand and is called the performance point.

Demand Spectrum Curve: Unfortunately, in Nepalese code, only design spectra are given. Hence, based on Indian code IS1893:2002, The demand spectrum curve for Nepal was plotted for different damping factor considering 0.36g PGA as maximum considered earthquake(MCE) of return period 500 years.

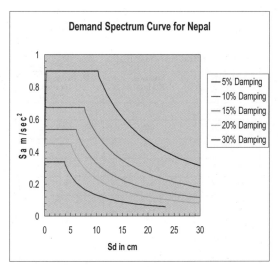

Figure 4 Demand Spectrum Curve

This simple static non-linear analysis method was used to determine the strength increasing factor (η). Firstly, the pushover analysis was carried out for the bare frame by applying incremental lateral loading according to the Ai distribution up to drift angle of 1/50 as the life safety limit. Similarly, analysis of the models by considering infill wall with first soft storey was carried out and it could not meet the demand. Then, the lateral strength of column in terms of size and reinforcement were increased gradually so that the strength could meet the demand. Therefore the strength increasing factor (η) is the ratio of the lateral strength of the strengthened frame structure subjected to soft storey to the lateral strength of the bare frame structure. Strength increasing factor (η) is function of initial storey stiffness ratio (Ko2/Ko1).

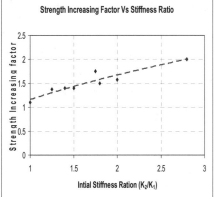

Figure 5 Capacity Spectrum Curves

Strength Increasing Factor (η) Vs Storey Stiffness Ratio

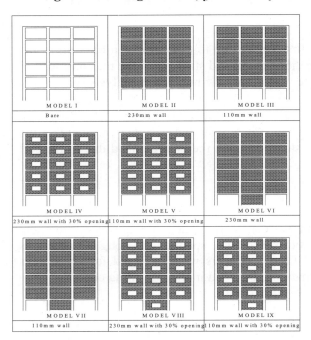

Figure 5 First soft storey models

As the soft storey means the abrupt change in stiffness along height of the building, the strength increasing factors depends upon storey initial stiffness ratio of the soft storey to the adjacent storey above it is a called initial stiffness ratio (Ko2/Ko1). Therefore, 2D models of six storied building with the most practical configurations of infill walls as shown in figure 5 were considered so that rational relationship could be established between Strength increasing factor (η) and initial stiffness ratio (Ko2/Ko1).

The pushover analysis for each of the models by considering the stiffness and strength of the infill wall (except Model I) was carried out in such a way that capacity curve of the all models would meet the demand by increasing its lateral capacity. The strength increasing factors were calculated for each model as a ratio of the lateral capacity of the strengthened to bare (ref figure.6). The strength increasing factor (η) with initial stiffness ratio (Ko2/Ko1) is tabulated in the table1

Table 1 Initial Stiffness Ratio

Model	Stiffness Ratio (K_1/K_2)	Strength Increasing Factor
Model I	1	1.1
Model II	2.8	2.0
Model III	2.0	1.56
Model IV	1.8	1.5
Model V	1.4	1.4
Model VI	1.75	1.75
Model VII	1.5	1.44
Model VIII	1.4	1.42
Model IX	1.25	1.36

Figure 7 Storey drift angle

Figure 8 η – ko2/ko1 curve

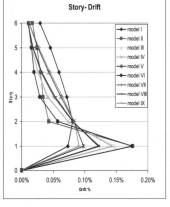

Result and Discussion:

The storey drift ratio for each models showed that the storey drift demand increased with the increase in the stiffness ratio (Ko2/Ko1). Hence, the relationship between strength increasing factor (η) and Initial stiffness ratio (Ko2/Ko1) could be established and plotted in the figure 7.

CASE STUDY: PRACTICAL VERIFICATION OF STRENGTH INCREASING FACTOR (η)

A typical six story high apartment typed building with masonry infill wall with open soft first storey was considered which is very common in the urban areas of Nepal as shown in figure 8. The plan of the building is symmetrical with both directions and grid of the column layout is 5m c/c in each direction and the storey height of the building is 3m.

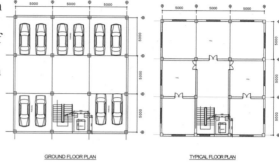

GROUND FLOOR PLAN   TYPICAL FLOOR PLAN

Figure 8 floor plans

Design of the Structural Elements

The structural design of the building was carried out to resist the seismic load as prescribed in Nepalese code.
Detailing of the members was done according to the Indian building code. The size of the columns were calculated as two type, peripheral columns were 500X500mm with 8# 20φ + 4# 25φ and similarly, inner columns were 500X500mm with 8# 25φ +4# 20φ. The 8φ ties were provided as 100c/c. Grade of concrete and reinforcement were 16MPa and 415Mpa respectively.

According to the graph of figure 8, the strength increasing factor (η) corresponding to Initial stiffness ratio 1.73 is 1.53. Hence, the columns of the building were redesigned for the base shear 1.53 times than that of the previous calculated base shear. The size and the reinforcement were calculated as 575 X 575mm and 12# 25φ respectively.

Table 2: Determination of η

Initial stiffness ratio		Strength increasing factor	
In X Dir.	In Y Dir.	In X Dir.	In Y Dir.
1.73	1.65	1.53	1.5

Non-linear Time History Dynamic Analysis

Non-linear dynamic time history analyses were carried out separately for bare, infill and strengthened infill frame with Kobe, Hachinohe and Chamauli (India).All the earthquakes were normalized to their maximum velocity of 50cm/sec. The Demand Spectrum for the response of all three normalized input ground motion were plotted in ADRS format. Similarly, the demand required by elastic response spectra was also plotted. Then, comparisons were done among them for all the three cases.

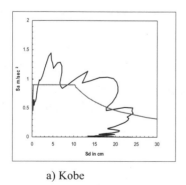
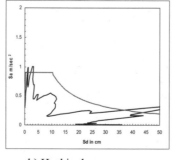
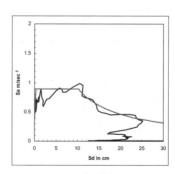

a) Kobe   b) Hachinohe   c) Chamauli (India)

Figure 9 Demand Spectrum Curves

Result and Discussion

All the three non linear dynamic time history analysis showed the development of soft storey in the first floor . By strengthening columns with the strength increasing factor (η), drift demand at first floor were considerably decreased. The result obtained from Chamauli earthquake which agrees with the demand spectrum showed that the drift ratio obtained after the strengthening the columns with strength increasing factor were almost 2% (Ultimate drift ratio). Hence, the proposed strength increased factor was well agreed.

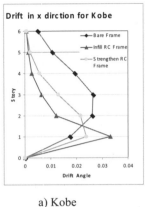
a) Kobe

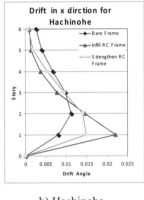
b) Hachinohe

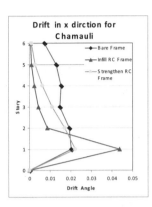
c) Chamauli (India)

Figure 10 Max storey drift ratio

CONCLUSIONS

- The effect of infill wall predominately changes the behavior of the structure and it is essential to consider infill walls for seismic evaluation of the structure.
- NBC105 underestimates the base shear by around 10%. Hence, the immediate review of the code is required.
- Arrangement of infill wall in the frame affects the behavior of the structure. A relationship can be established between strength increasing factor(η) and initial stiffness ratio. Designer can use strength increasing factor (η) to take care of soft storey effect without carrying out non-linear analysis
- The non-linear dynamic time history analysis for 3d models proved that the demand of the soft first storey can be met by application of strength increasing factor (η).

ACKNOWLEGDEMENT

I would like to extend my sincere gratitude to my advisor Dr. Taiki Saito, the chief researcher and Engineering course leader of IISEE for letting me to use his software STERA3d, continuously supporting and encouraging me during my study.

REFERENCES

A. Madan,A.M. Reinhorn, ,J. B. Mandar, R.E. Valles, 1997, ASCE, Vol.114, No.8, pp.1827-1849
T. Pauley, M.J.N. Pristley, 1992, Seismic Design of Reinforced Concrete and Masonry building, JOHN WILEY & SONS, INC
ATC, Seismic Evaluation and retrofit of concrete buildings, Rep ATC 40, Applied Techonology Council, Redwood City, California (1996)
Dr. T. Saito, 2005, Structural Earthquake Response Analysis 3D (STERA 3D Ver. 4.2) Manual, Building Research Institute, Japan
NBC-105, 2004, Nepal National Building Code for Seismic Design of Buildings in Nepal,

Synopsis of Master Papers   *Bulletin of IISEE, **43**, 91-96, 2009*

EARTHQUAKE RISK PERCEPTION OF STAKE HOLDERS INVOLVED IN HOUSING SAFETY IN PAKISTAN

Ghazala Naeem[*]   Supervisor: Dr. Kenji Okazaki[**]
MEE07155

ABSTRACT

Pakistan is vulnerable against potential seismic risk and has recently suffered from the disastrous earthquake in October 2005 causing enormous human and economics losses. In this study, after an outline of seismicity of Pakistan, damages by the 2005 Kashmir earthquake and consequent reconstruction activities, results and analysis of a survey conducted among all stake holders involved in housing construction are presented. The survey consisted of various questions on seismic risk perception of house owners (residents), house builders/ head masons, central and local government officers. As resident's survey was conducted in two different communities, one was severely affected and other was not damaged by the 2005 Kashmir Earthquake, their responses are different about earthquake risk perception but similar for mitigation efforts. Some similar outcomes were found in case of survey analysis of builders/ masons, central and local government officers. The 2005 Kashmir Earthquake has triggered the consciousness among all stake holders in Pakistan especially people who have either suffered the disastrous consequences of the earthquake or have been involved in post earthquake relief and rehabilitation activities in affected areas. This potential needs to be utilized in an optimum way as such memories can be faded away quickly.

Some suggestions, to improve housing safety based on the analysis are proposed to raise disaster mitigation awareness and disseminate information regarding safer technologies effectively, to communities. This study may be helpful while planning mitigation programs and public awareness campaigns on preparedness in Pakistan and areas with similar characteristics.

Key words: Earthquake risk perception, housing safety, disaster reduction

BACKGROUND AND METHEDOLOGY

The 2005 Kashmir Earthquake demonstrated the extent of damage, an earthquake can cause in Pakistan. Given to Pakistan's seismo-tectonic setting, this earthquake is not a one time event, but a part of sequence of earthquakes that happened in the past and will happen in the future. Most seismically active areas are north, northwestern and western sections of the country along the boundary of the Indian Tectonic Plate with Iranian and Afghan Micro Plates. The number of fatalities and injuries exceeds 73,000 and 125,000, making it by far the most fatal earthquake ever to occur in the Indian subcontinent or its surrounding plate boundaries. Most of the deaths were caused by the collapse of buildings that were not adequately designed for earthquake resistance,

---

[*] Earthquake Reconstruction and Rehabilitation Authority, Pakistan
[**] Professor, National Graduate Institute for Policy Studies, Japan

were poorly constructed using stone, fired brick or concrete blocks using mud or cement mortar. In 1935 a Richter magnitude M7.5 strike-slip earthquake near the city of Quetta (the only large settlement in an otherwise sparsely populated region between Afghanistan and Pakistan) resulted in an estimated 35,000 dead.

In Pakistan mostly buildings have been procured by the owners themselves by employing unqualified and unskilled contractors/ masons for saving cost as for ordinary private construction there is no legal frame work for licensing requirement of contractor. It is evident that there has never existed an effective building monitoring mechanism. Even in cities municipal organizations do not have the institutional capacity for the strict implementation of the code for building construction, making almost all the building stock inappropriate especially, for a seismic region. .

The 2005 Kashmir Earthquake proved the vulnerability of housing stock and fatal consequences which lead to a realization of safer houses in Pakistan. During past three years of reconstruction activities in earthquake affected areas, national and international organizations are making efforts to ensure safety of post earthquake construction of the affected areas. Therefore, this study aims to analyze hazard-related human behaviors to identify the factors that determine the understanding and interpretation of the people: how the people perceive seismic risks, how such risk perception would be biased by economic and social aspects, how they would like to avoid such risk, using social survey data carried out in Pakistan, involving key players for earthquake safety i.e. residents, government officials and house builders / head masons.

This survey was conducted in 2007 as a part of Collaborative Research and Development Project for Disaster Mitigation under the program of Building Research Institute (BRI), Japan and National Graduate Institute for Policy Studies (GRIPS) Japan, using the questionnaires developed by Dr. Kenji Okazaki (GRIPS). All interviews were carried out by visiting interviewees in their houses and work places through face to face communication and questionnaires were filled by the surveyors according to the answer of the respondents. As there has been no such survey on earthquake risk perception of communities and other stake holders concerned with housing safety in Pakistan, outcome of the research will be beneficial to the experts and policy makers while planning mitigation programs and public awareness campaigns on preparedness and mitigation for the regions with similar characteristics.

SURVEY DATA AND ANALYSIS

Residents

A total of 800 households were surveyed from two different communities each represented by a village having distinct characteristics, one lies in the high seismic zone severely affected by the 2005 Kashmir Earthquake (village Panyali, District Bagh) and the second lies in the zone which has a very low seismic activity and has practically seen no earthquake activity (village Kamman, District Okara). This choice of two regions separated by distance and experience of disasters was to study the perception of the two people about disaster in general and earthquake disasters in particular. Respondents were asked questions to know how they perceive seismic risk like: whether they think their house is safe against earthquakes, how they want to avoid the risk of damage to their house and to their family, what they know about retrofitting etc., in addition to the questions about their sex, age, number of family members, household income, occupation, academic qualification, and house related information such as floor area, type of house, cost and ownership.

In both communities, 96% respondents were males because of socio-cultural restrictions imposed on women's social participation and mostly respondents belong to age group between 30-60 years. Most of the total population (85%) has either school education or cannot read /write at all, although in Panyali academic qualification level is higher than Kammna and comparatively better occupations. Average house hold size in Panyali and Kamman are 6.4 and 7.8 respectively. Almost all houses are self/ family owned having floor area more than 200 sq m because of living styles and type of houses. Respondents in Panyali are staying longer period in the same house i.e. more than 50% are living between 25 and 50 years in the same house. In Kamman almost 70% people are living less than 25 years period. In terms of house type, both communities have independent houses because of socio-cultural trends and environmental requirement in Pakistan. There is a general trend in Pakistan of building the house by hiring mason or a local builder (labor contractors); therefore both surveyed communities mostly have self built houses built by the local mason / house builders (85%-96%). Cost comparison of self built houses shows Panyali has a larger percentage of house built with less cost than in Kamman. Uuse of locally available stones with mud mortar is a major construction material in Panyali, resulting low cost of construction as compared to plain areas like Kamman where mostly houses are built with burnt brick using cement mortar.

In response to the question about any disaster experienced in life before, almost respondents in Panyali have experienced recent 2005 Kashmir Earthquake whereas in Kamman 73% have experienced flood because of being flood prone region. Majority of respondents in Panyali (76%) consider disaster as most affecting event because all of them have experienced the 2005 Kashmir Earthquake and its aftermaths where as in Kamman 33% considers unemployment, 27 % disaster and 25% disease. Overall 84%-88% people think that a future earthquake can cause loss of lives, injuries and loss of properties as well. In both communities almost all respondents consider their houses unsafe against a big earthquake. More than half of the Panyali respondent relies on masons/carpenters whereas almost same percentage in Kamman depends on government for safer construction. Almost 50% of overall population considers use of poor construction materials/ works as the major cause of the house collapse in case of a big earthquake. It seems that mostly Panyali residents are aware of the technical flaws about weak construction of their houses which had caused swerve damage to life and property by the 2005 Kashmir Earthquake. Mostly people (96%) are unaware of the available techniques of strengthening their houses. Similarly, respondents have estimation for cost of strengthening the house against earthquake either equal or even many times the cost of self built house within groups of low cost houses which seem to be unrealistic. On the average almost 80% respondents are ready to spend more than five years of their income to protect their house/ property from big earthquake. Almost 40% Panyali respondent have plans either to build/ purchase an earthquake resistant house or strengthen their existing house by retrofitting. In Kamman this percentage is relatively lower i.e. 32.7%. A large portion of the Kamman population (40%) has all together no plan for safer house but in Panyali only 2.3% people fall in this category. A great majority of Panyali residents (63%) are aware of the organization working for disaster risk reduction in their area and mostly they have participated in community based disaster reduction activities as compared to Kamman respondents who mostly don't know about any such organization. The significant difference in responses can be attributed to ongoing reconstruction and mitigation activities in earthquake affected area therefore mostly Panyali residents have participated in community activities for disaster risk reduction.

Findings

For most affecting event, disasters are considered to be the priority event by most of the residents but it can be noticed, there is a general high tendency of this response with the increase in income and academic qualification levels (Figure 1 & 2). Earthquake is considered to be the most affecting disaster within high income and academic qualification groups as seen in Figure 3 & 4. Similarly, people having high income level and school, college and university education have relatively better tendency towards retrofitting the house, considering their present house not strong enough to resist big earthquakes (Figure 5 & 6).

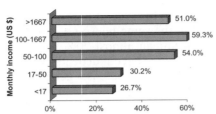

Figure 1: Monthly income disaster as most affecting event

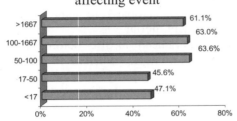

Figure 3: Monthly income and earthquake as the most affecting disaster

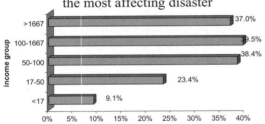

Figure 5: Relation between income group and respondent who have retrofitted the house

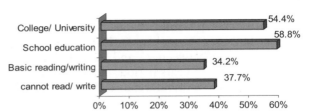

Figure2: Academic qualification & disaster as the most affecting event

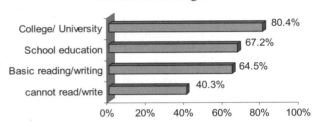

Figure 4: Academic qualification and earthquake as the most affecting disaster

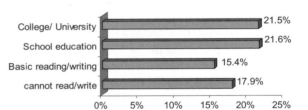

Figure 6: Academic qualification and retrofitting as the plan for safer house

Composition of respondents considering strengthening cost of their houses ranging between US $ 3333-8333 and more in relation to cost of self built house is presented in Figure 7 which shows that tendency of estimated strengthening cost more than US$ 3280 is more within group with high cost of self built house. Although in most cases, estimated such cost is either equal or more than the actual cost of the respondent's house even less initial cost.

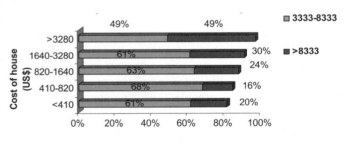

Figure 7: Cost of house and expected cost of retrofitting

House Builders/ Head Masons

Total of 37 house builders and 13 head masons were interviewed from Islamabad, Lahore, Faisalabad, Multan, Sahiwal and some interior villages of Punjab and Northwestern Frontier province working in various companies and groups. Only two builders were qualified engineers and the rest have ten years of school education whereas most of masons have basic reading and writing skills. Mostly builders/ head masons (94%) provide services either on labor contract or labor and material contract. Builders and masons who think that most contributing factors to loss of lives in case of earthquake would be house collapse is almost 90%. Response to this question shows the awareness of the builders about the vulnerability of dwellings people are living in. Only 4 % builders/ masons have received any formal training about earthquake resistant construction because in Pakistan there is no system or law of licensing, registration or training of builders/ head masons. Similarly, about using building code or construction guidelines, 28 % builders/ head masons have been applying its provisions in building construction and the rest 72 % have either never heard about such details or just heard but don't know details. There are 4% of builders/ masons who know these details but not using in construction considering impractical. It is important to mention that for ordinary house construction there is no legislation for application and adoption of building code neither in design nor in construction practices. Majority of respondent (38%) expect government to develop a proper licensing system and carry out training programs on earthquake resistant construction, to ensure rained masons. One of the major difficulties that 17 builders/ masons mentioned is homeowners resist as they don't want to pay extra cost for making an earthquake resistant house. Almost 38 % of overall respondents consider retrofitting feasible; on the other hand 42 % are those who either consider retrofitting impossible or not viable. The rest 20% don't know about retrofitting at all. These responses give a clear picture of the lack of awareness of builders/ masons about retrofitting who are responsible for construction of 90 % houses.

Central Government Officers

Total of eight officers were interviewed who are well aware of the aftermaths of the 2005 Kashmir Earthquake as most of them are directly or indirectly involved in reconstruction and disaster management activities or policy making. Almost half of the respondents anticipate immediate threat whereas the others either don't know or do not anticipate a near future earthquake. Except two officers who do not have any idea about highest anticipated loss of lives all others anticipate casualties in millions and thousands, in case of a future earthquake. As majority considers poor construction of buildings as the most contributing underlying factor to earthquake disasters in Pakistan and the main causes for vulnerable building stock are ranked as lack of building enforcement system, economic conditions of people leading not to able to afford good material and technology, lack of awareness among public and lack of appropriate technical know-how. Considering the available resources/capacity and prevailing risk situation four of the officers prioritize need for building code enforcement as the pre-disaster measures for the earthquake safety whereas the rest give priority to other risk management policies. Further, they recommend strong legislation, public awareness scheme, training of building control staff and financial support to house owners for effective implementation of building control system in the country.

Local Government Officers

Twenty five Local government officers working in different cities and districts of province Punjab, Azad Jammu and Kashmir and Northwestern Frontier Province, have been interviewed. Officers working in Punjab face flood as frequent disaster in the area of their domain whereas others have experienced the 2005 Kashmir Earthquake and/or participated in relief and reconstruction activities. Local government officers who expect a big earthquake within few years and within ten years constitute 48% and all officers anticipate large casualty and buildings collapse in case a big earthquake hit the area. Two third of the respondents believe that vulnerable building stock due to bad construction practices, is the most contributing earthquake risk factor to city. Almost 20% considers lack of mitigation efforts and 8% opted for the vulnerable life line structures, in this regard. Further majority of the officers (56%) believe that homeowners cannot afford to have good materials and utilize earthquake resistant technologies, which is the most critical cause for vulnerable building stock. Whereas 28% believe, lack of appropriate technical know how and inaccessibility to such techniques of the house owners are the critical causes for vulnerable buildings. All officers believe that a building permit system exists in their cities principally and building code implementation is supposed to be a part of that system but unwillingness of the public to follow the building code is the major issue for effective implementation of the building code. Most of them believe that builders/petty contractors and masons can contribute more to improve building safety as most of the housing stock is constructed by them.

CONCLUSIONS

The study has revealed a clear difference in perceptions about future earthquake risks and minor differences in mitigation efforts done so far or planned to do in future, between people of the two areas perception; one which was hit by the earthquake having relatively high risk perception as compared to other which is quite far from the earthquake prone area. Similarly finding of the survey shows there are different level of understandings of people belonging different income groups, academic qualification and disaster experience. Therefore, objectives of awareness raising should differentiate between target groups focusing their understanding level, role and importance in housing construction, affordability, accessibility and acceptability. The 2005 Kashmir Earthquake has created a consciousness among all stake holders involved in housing construction in Pakistan. This potential needs to be utilized in an optimum way as people can forget such experiences very quickly. Therefore it is necessary to preserve data, facts and lesson learnt by the earthquake to communicate and disseminate knowledge and wisdom gained with communities. For this purpose, an institute is proposed which can share the effects of the 2005 Kashmir Earthquake and promote disaster risk mitigation within communities following the pattern of Disaster Reduction Institute (DRI), Kobe which offers programs by which visitors can learn the effects of the Great Hanshin-Awaji Earthquake and lessons learned from the experience that should be shared with younger generations. This gives inspiration to the visitors for realizing the disastrous impacts of earthquake and motivation to adopt measure for safer environments. Further, DRI is involved in research and training programs promoting social preparedness against possible major disasters in and outside Japan. These conclusions may be very helpful in planning mitigation programs and public awareness campaigns in Pakistan and regions with similar characteristics as well.

ACKNOWLEDGMENT

I would like to express my sincere gratitude to Mr. Masahiko Murata (DRI, Japan) and Mr. Najib Ahmed (Preston University, Pakistan) for providing useful data, support and cooperation during my individual study.

REFERENCES

Kenji Okazaki, 2008, Study on Risk Perception concerning Housing Safety agaianst Earthquake-EAROPH , Seismic Risk Perception of People for Safer Housing, 14th World Conference on Earthquake Engineering, 2008, Beijing, China

Najib Ahmed, 2007, Report on Disater Risk Reduction, Preston University
Websites, www.nha.gov.pk, http://web.worldbank.org/, , www.dri.net.jp, www.erra.gov.pk,

Synopsis of Master Papers   *Bulletin of IISEE, **43**, 97-102, 2009*

RESPONSE CONTROL BY DAMPER DEVICES OF HIGH-RISE BUILDING UNDER LONG-PERIOD GROUND MOTION

Miguel Augusto DIAZ FIGUEROA[1]   **Supervisor: Taiki Saito**[2]
MEE07150

ABSTRACT

This study attempts to apply Performance Curve method to design dampers for high-rise buildings in order to reduce the response under long-period ground motions; and to analyze the effects of response control of the high-rise buildings using damper devices.

In order to analyze the dynamic response, the target high-rise building is a 37-story building located in Tokyo; and five different earthquakes, namely BCJ-L2, El Centro 50 kine, Kobe, Osaka and Nagoya are used. Likewise, the target high-rise building is modeled as multi-degree of freedom (MDOF) systems, and to perform its dynamic response using damper devices, such as steel, oil and viscous dampers, the software called MDOF-OS was modified by the writer based on the software coded by Dr. Taiki Saito. Its algorithm is mainly based on Operator Splitting (OS) method because of its good accuracy and stability for large time integration steps. In this analysis, the target high-rise building without dampers reaches the maximum response under Nagoya earthquake because of being the strongest long-period ground motion.

Finally, steel and oil dampers are designed using a Performance Curve method developed by Prof. Kasai to be added into the target high-rise building in order to reach a good seismic performance under Nagoya earthquake. And a comparison among these dampers is carried out in order to analyze the effect of response control.

Keywords: High-rise building, long-Period ground motion, passive control, Performance curve method.

INTRODUCTION

In the last decades, the response of high-rise buildings caused by long-period ground motions has been a subject of research, considering that responses of structures with long natural period, such as high-rise buildings become very large in case of long-period ground motions; it is desirable to control the response of these structures to avoid damages and insure the life safety. In that sense, the response control must especially be applied on existing high-rise buildings, and these measures must also be taken in account in the design of new high-rise buildings.

In order to control the response of these structures, many types of passive dampers devices were developed in Japan, these devices are connected to the frame to dissipate the seismic input energy, therefore reducing the kinetic energy and vibration of the building.

Prof. Kasai developed and proposed a simple and useful design procedure for buildings with dampers, so-called Performance Curve method, which is described in the Manual for Design and Construction of Passively-Controlled Buildings published by Japan Society of Seismic Isolation (JSSI). However, few researches of application of response control devices to high-rise building to solve the aforementioned problem of high-rise buildings under long-period ground motions were carried out.

---

[1] Japan Peru Center for Earthquake Engineering and Disaster Mitigation (CISMID)
[2] Chief Research Engineer of Building Research Institute (BRI)

DATA

Target High-rise Building

The target high-rise building is located in Chuo-ku, Tokyo; and its construction end was in 1991. The total height is 119 m, compose by 40 floors, as is shown in Figure 1. In order to model this building, one basement and two minor upper floors are neglected, so that it is supposed as 37-story building, as is shown in Figure 3, with a total height of 108.9 m. On the other hand, this structure is symmetric in plan as is shown in Figure 2.

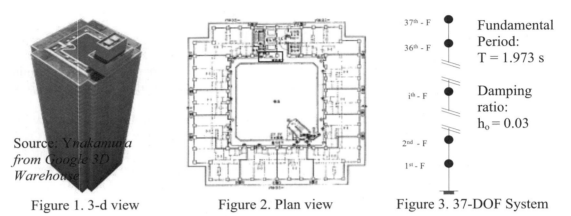

| Figure 1. 3-d view | Figure 2. Plan view | Figure 3. 37-DOF System |

Source: *Ynakamura from Google 3D Warehouse*

37th - F
36th - F
ith - F
2nd - F
1st - F

Fundamental Period:
T = 1.973 s

Damping ratio:
$h_o = 0.03$

Input Earthquake Ground Motions

Table 1. Summary of used Input Earthquake ground motions

Earthquake	Type earthquake	Characteristics
BCJ-L2	Design earthquake	Building Center of Japan (Level 2). Artificial wave.
El Centro 50 kine	Design earthquake	1940 Centro Earthquake, NS component. Amplified to obtain a maximum velocity of 50 cm/s.
Kobe	Observed earthquake	1995 Hyogoken Nanbu Earthquake, NS component.
Osaka	Long-period ground motion	An artificial wave set in NS component, simulated for Nankai Earthquake.
Nagoya	Long-period ground motion	Artificial wave set in EW component, simulated for Tokai-Tonankai Earthquake.

Since this study attempts to analyze the effect of damper devices into the target high-rise building, Nagoya earthquake is selected because of being the strongest long-period ground motion.

Long-period ground motions greatly influence structures with long fundamental period, approximately from 2 up to 6 s, as is shown in velocity spectra comparison as is shown in Figure 4.

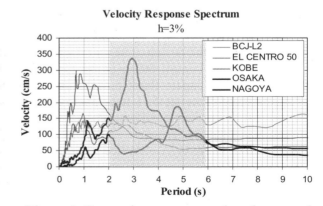

Figure 4. Comparison among earthquake ground motions

THEORY AND METHODOLOGY

Performance Curve Method

A damping structure controls any displacement, velocity and acceleration response of a building by means of stiffness and viscosity added by dampers attached to the main resisting frame when an earthquake occurs. The performance curves are the visualized response conditions in a single-mass system. The frame stiffness has been used as a reference for all the stiffness values and to plot the performance curves as is shown in Figure 5.

The displacement reduction ratio, R_d, is the initial parameter to find out the optimum amount of dampers into the performance curve and it is defined by the following Equation:

$$R_d = \frac{\text{Maximum target drift}}{\text{Maximum response drift}}$$

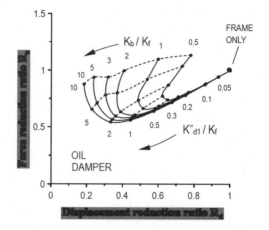

Figure 5. Performance curve

Design of dampers

A general procedure to design dampers is described as follows:

- Target performance and design conditions.
- Calculating the displacement reduction ratio.
- Plotting the damping performance curve.
- Determining the required damper scales for each story. The amount of dampers from the performance curve is obtained to equivalent single degree of freedom system; therefore the actual amount of dampers is obtained by means of a linear proportional distribution along MDOF system.
- Converting to the damper axial values and determining the specifications. The MDOF system assumes a horizontal displacement of dampers; thereby a conversion of forces and displacements from horizontal to diagonal direction using the incidence angle is done in order to design dampers at each floor.
- Designing dampers.

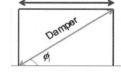

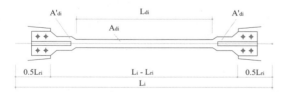

Figure 6. Steel damper used in this study. (Buckling Restrained Damper)

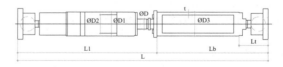

Figure 7. Oil damper device used in this study.

RESULTS AND DISCUSSION

Dynamic Response of the Target High-rise Building without Damper Devices

The maximum values of dynamic response under the above earthquake ground motions are shown in Figure 8 and 9, and summarized in Table 6 which shows that the maximum response is obtained under Nagoya earthquake, being the maximum drift 1/80 rad at the 14-floor and the ductility factor is 1.594.

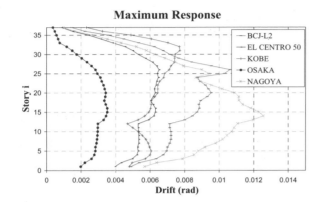

Figure 8. Maximum Story drifts   Figure 9. Maximum Story shear forces

Table 2. Maximum values of dynamic response under earthquake ground motions

Earthquake ground motion	Base shear force *t*	Maximum drift *rad*	Ductility (μ)
BCJ-L2	6400	1/94 at 26-F	1.346
El Centro 50	5520	1/153 at 26-F	0.836
Kobe	6400	1/129 at 31-F	1.195
Nagoya	6400	1/80 at 14-F	1.594
Osaka	3290	1/284 at 15-F	0.423

Design of dampers

The maximum story drift for designing high-rise buildings is 1/100 rad, and the obtained maximum story drift under Nagoya earthquake is 1/80 rad; therefore the reduction ratio, R_d, is 0.80. The obtained results for steel and oil dampers are shown in Table 3 and 4, respectively.

Steel dampers

Table 3. Specifications of steel dampers devices used in the design

Type	L mm	Lr mm	Ld mm	Plate	Ad cm^2	A'd cm^2
A	4066	407	3253	PL-22x250	55.0	137.5
B	4066	407	3253	PL-28x300	84.0	210.0
C	4066	407	3253	PL-30x320	96.0	240.0
D	4066	407	3253	PL-32x380	121.6	304.0
E	4066	407	3253	PL-38x400	152.0	380.0
E'	5327	533	4262	PL-38x400	152.0	380.0

Oil dampers

Table 4. Specifications of oil dampers used in the design

Type	Fdy KN	Cd KN.s/cm	Vd cm/s	L mm	L1 mm	Lb mm
A	1500	1350	900	4066	3000	1066
B	1000	900	300	4066	3000	1066
C	1500	1350	900	5327	3000	2327
D	1000	900	300	5327	3000	2327

The design of steel and oil damper was carried out and the arrangements of dampers at each floor are shown in Figure 10 and 11, respectively.

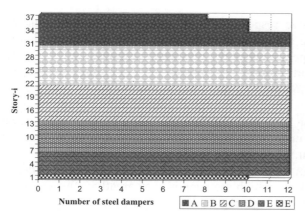

Figure 10. Arrangement of steel dampers

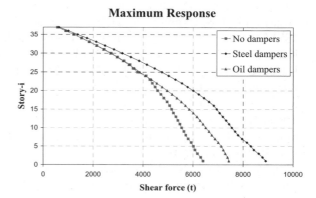

Figure 11. Arrangement of oil dampers

Dynamic Response of the Target High-rise Building with Damper Devices

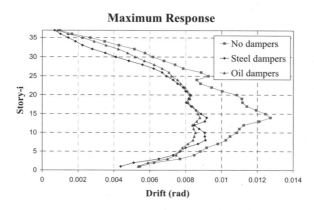

Figure 12. Comparison among maximum drifts

Figure 13. Comparison among maximum forces

As was supposed, the adding of dampers reduces considerably the dynamic response under Nagoya earthquake (the strongest long-period ground motion) in both cases, adding steel and oil dampers. The obtained maximum drift at the fourteen-floor is reduced closely to the target maximum drift, these differences are in 8% and 10% under the maximum target drift, 1/100 rad, as is shown in Table 5, with the adding of steel and oil dampers, respectively; these differences demonstrate the high control on buildings. Also, ductility factors are reduced in 19% and 24% adding steel and oil dampers, respectively.

Table 5. Summary of maximum response before and after adding dampers

Condition	Base shear-force t	Maximum drift rad	Ductility (μ)
No dampers	6400	1/80 at 14-F	1.594
Steel dampers	8910	1/109 at 14-F	1.293
Oil dampers	7430	1/114 at 14-F	1.216

CONCLUSION

In this study, an analytical investigation on passive response control using steel, oil and viscous dampers has been presented in order to analyze their effects on an actual high-rise building in Tokyo.

The target high-rise building without dampers reaches the maximum response under Nagoya earthquake because of being long-period ground motion. It verifies that buildings with long fundamental period, approximately from 2 up to 6 seconds, are greatly influenced by long-period ground motions.

The design of dampers was carried out for Steel and Oil dampers using Performance Curve method developed by Prof. Kasai. The responses under long-period ground motion after adding the above dampers satisfy the design safety requirement.

This procedure can be applied in general for high-rise buildings to design dampers in order to reduce the response under long-period ground motions, which are supposed to greatly influence these structures. This allows us avoiding severe damaged on high-rise buildings and insure the life safety, which are the desired targets in the design.

REFERENCES

Chopra A., 2007. Prentice Hall, Inc., New Jersey.
Higashino M. & Okamoto S., 2006. Taylor and Francis Group, New York.
JSSI Manual, 2004. Japan Society of Seismic Isolation (JSSI), Tokyo (in Japanese).
Kasai K., Fu Y. & Watanabe A., 1998. Journal of Structural Engineering.
Kasai K. & Kibayashi M., 2004. Proc., 13WCEE.
Kibayashi M., Kasai K. &, Tsuji Y., Kikuchi M., Kimura Y., Kobayashi T., Nakamura J. and Matsubsa Y., 2004. Proc., 13WCEE.
Tada M. & Pan P., 2006. International Journal for Numerical Methods in Engineering 2006.
Soda S., 2008. Lecture notes of International Institute of Seismology and Earthquake Engineering, Building Research Institute.
Iiba M. & Inoue N., 2008. Lecture notes of International Institute of Seismology and Earthquake Engineering, Building Research Institute.

PERFORMANCE BASED EARTHQUAKE EVALUATION OF SCHOOL BUILDING IN SRILANKA

Chandima Kularatne*   **Supervisor: Dr. Tomohisa Mukai****
MEE07153

ABSTRACT

This study quantifies the seismic vulnerability of typical three storied class room building located in SriLanka. A significant feature of this reinforced concrete frame building is that this was not designed for seismic effects. Besides, there are some infill brick walls also.
A methodology given in "Guidelines for Seismic Performance Assessment of Buildings"(ATC-58) was followed as it was expected to study consequences of earthquake. As per ATC-58, intensity based assessment was considered for Hazard Analysis while the Structural Analysis was carried out to obtain engineering demand parameters (EDP) such as story drift and story acceleration. Besides, a Damage Analysis was carried out not only for structural members but also for non-structural members. Pushover analysis was done in order to define damage states of the structural while Loss Analysis was carried out to quantify the direct economic loss.

1.INTRODUCTION

School buildings have been considered as post-disaster buildings or special buildings in many countries considering its' role after an earthquake. As per HAZUS-MH MR3 Technical Manual also, school buildings are considered as essential facilities. Therefore, it is it is very important to design school buildings to withstand seismic effects.

However it has been realized that yet non of the school buildings in SriLanka have been designed to mitigate seismic effects. Therefore, it is important to investigate the seismic vulnerability of at least a single school building in SriLanka. A three storied building was selected for investigations because this is the most typical building all over the country. In order to achieve above objectives, a methodology given in "Guidelines for Seismic Performance Assessment of Buildings" (ATC-58) developed by Advanced Technology Council was used. Accordingly, an assessment of performance capability was carried out for this building.

The software STERA 3D (Dr.T.Saito,2005) was used to perform pushover analysis and nonlinear response-history analysis. Pushover analyses were used to define the story capacity limit states of the modeled structure. Pushover analyses were run several times to determine quantitative limit states of inter-story drift. Nonlinear response-history analysis was used to determine the engineering demand parameters. A simple calculation method was used for loss calculations. The repair method was proposed as per in the context of SriLanka and the prevailing rates were used for cost calculation.

---

* Civil Engineer(Structural Designs),Ministry of Education, Government of Sri Lanka.
** Research Engineer at International Institute of Seismology and Earthquake Engineering

2.BUILDING OVERVIEW

Even though this building has been designed for wind loads, it can be proved that the wind load is very small compared to the seismic load. The designed strength of concrete has been considered as 20MPa. Since this is cube strength of concrete, it was converted into cylindrical strength as 16Mpa. The strength of all the main reinforcement steel is 460Mpa while the strength of shear reinforcement is 250Mpa. Besides, the compressive strength of bricks has been estimated as 3Mpa as quality of the bricks are poor. The compressive strength of mortar has been considered as 4Mpa. The detail of this building is shown in Figure 1. A size of 300mmX375mm columns are located from first floor to third floor level and corner columns from third floor to roof level. Size of the other columns from third floor to roof level are 225mmX375mm. Size of all the beams spanning in longer direction of the building is 225mmX22mm. However, the size of beams spanning shorter direction is 300mmX650mm except roof level. The size of beams spanning in shorter direction at roof level is 225mmX450mm. Besides, there are some masonry infill walls spanning in the shorter direction. Here after, the longer direction of this building was considered as X-X direction while the other direction was considered as Y-Y direction.

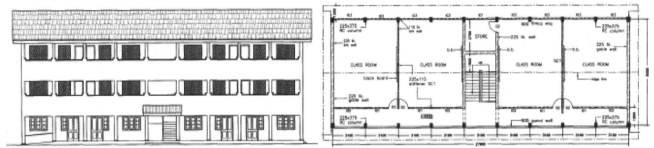

Figure1. Detail of the building

3.HAZARD ANALYSIS

As we are aware, earthquakes can also occur within the single tectonic plate and are referred to as intraplate earthquakes. SriLanka is also located within one of such area know as Indo-Australia plate. Intraplate regions have been historically described as areas of low to moderate seismic hazard. The effects and risks involved with intraplate earthquakes have been studied, but not yet fully understood. Especially in SriLanka, lack of reliable data on past earthquakes hinders the study on seismicity around SriLanka. However, as per available records, it is evident that SriLanka is not a seismically inactive country but the earthquake could occur at long recurrence period

According to the available data, it was suggested for Colombo an earthquake of magnitude close to $M_L=6$ on Richter scale with a return period of 200-400 years; a design acceleration of $0.2g(196cm/s^2)$ is considered as the horizontal component of the earthquake while the vertical component is neglected at this stage. Besides, the attenuation relation developed by Fukushima and Tanaka was applied for earthquake of $M_L=6$ on Richter scale and a site of 5km away from the epicenter. It was found that the Peak Ground Acceleration as 186.4 cm/s^2 which is quite similar to suggested acceleration 0f 0.2g. Therefore, it was decided to follow intensity based assessment with a acceleration of 0.2g ($196cm/s^2$).

4.STRUCTURAL ANALYSIS

Since this is an existing structure, structural designs were not carried out. However, the failure modes of the building were investigated and it was revealed that all the beams and columns failed because of flexure. Besides, it was revealed that the failure modes of the columns and beams are columns governed by the flexural beam and flexural column in X-X and Y-Y directions respectively.

Displacement control pushover analysis was used for this study. First of all, a target displacement for the structure was established as 1/50. The contribution of the infill walls to seismic resistance was also considered in this study. The effect of P-Delta was not considered for the analysis because of the low rise building.

According to pushover analysis, it was revealed that the soft-first story mechanism would be formed in Y-Y direction and more interestingly it was observed that the obtained failure modes were agreed with that obtained from manual calculations

Since it was observed several infill masonry walls, an attempt was made to develop an analytical model. The shear resistance at cracking, maximum and ultimate was calculated as per Kabeysawa, T., Mostafei, H. It was noticed that the respective displacements are almost similar in analytical model and results of pushover analyses.

Nonlinear Response-History Analysis was used to determine the engineering demand parameters(EDP) of the structure with 5% damping under suite of 11earthquake ground motions. The suite of earthquakes was scaled to spectral acceleration of 196gal at the period of 0.903sec and 0.522sec because it was expected to analyze the structure for both directions.

5.DAMAGE ANALYSIS

According to ATC-58, it is necessary to assess the possible distribution of damage to structural and nonstructural components using the response data from the structural analysis together with data on the building configuration. However, the damage analysis of this study was concentrated mainly on structural components and masonry infill walls.

Accordingly the structural damage states such as slight, extensive and complete of structural members and masonry walls were classified based on the pushover analysis results. The damage states are shown in Figure.2. Besides, the repair methods for each damage state were studied.

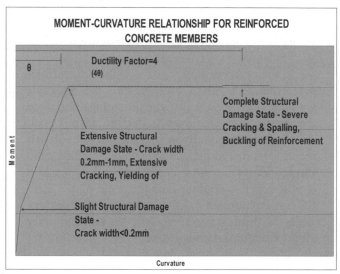 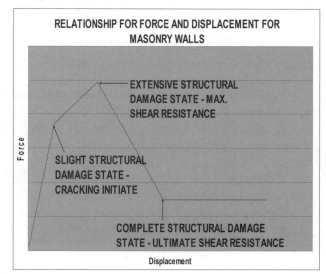

Figure 2. Damage states of structural members and infill masonry wall

It was observed that all most all of the beams spanning in X-X direction reached the complete damage state in case of pushover analysis was performed in X-X direction. Thus, it was determined that the base shear coefficient of this structure at X-X direction is 0.08 and this is not sufficient even for regions where seismicity is very low. Besides, story drift at respective damage state was almost same for second and third story. However, it was somewhat different in roof level and out of thirty six(36) beams spanning X-X direction, only twenty beams reached complete damage state.

Similarly, pushover analysis results on Y-Y direction were also studied. It was found that only some of the columns reached these damage states. The beams spanning Y-Y directions did not reach even the slight structural damage state. Even at second story only sixteen columns had reached the complete structural damage stage. However, it was found that the base shear coefficient of Y-Y direction is 0.245, as such that is sufficient enough for low seismicity countries like SriLanka.

In addition, it was decided to define structural performance level of this building based on Prestandard and Commentary for the Seismic Rehabilitation of Building published by FEMA(FEMA 356). Therefore, an attempt was made to represent the probable Structural Performance Levels of this building in terms of top drift. According to the results of pushover analysis, the relationship of top drift versus base shear coefficient for both directions were drawn and shown in Figure 3.

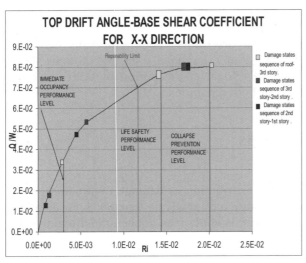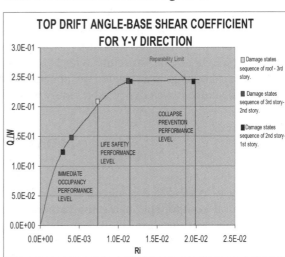

Figure 3. Damage sequence of members X-X and Y-Y direction

In addition to concrete frame structure, behavior of the masonry infill walls were studied according to the above defined performance level. Results of pushover analysis were used to identify the top drift when masonry walls reached respective performance level and which is depicted in Figure 31.

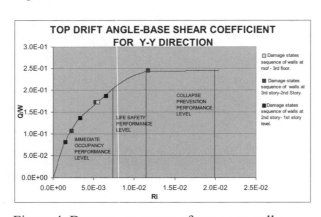

According to the defined performance level of building, all the masonry walls at first story level will be collapsed before the Immediate Occupancy Performance Level. All the masonry walls at second story level will reach extensive damage stage before the Immediate Occupancy Performance Level while it will reach complete damage state at the Life Safety Performance Level. However, masonry walls at third story level will be reached only slight structural damage state.

Figure 4. Damage sequence of masonry walls

Subsequently, the engineering demand parameters (EDP) obtained from nonlinear response-history analysis were used to study the vulnerability of this structure. It was proven that the X-X direction of the building is the most vulnerable. It was observed that the maximum response story drift of the X-X direction in the second floor did not exceeded the life safety performance level for the suite of earthquake. However, it was not exceeded the immediate occupancy performance level in case of maximum response story drift for the suite of earthquakes on Y-Y direction and not even all the columns in the first story level did reach the extensive structural damage state. It was found that the behavior of the second story level is similar to that of first story level.

It was observed that the performance of the third story in both directions is significantly better than that of first and second story levels. The both, maximum response story drift and minimum median response story drift in X-X direction was lain within the life safety performance level while that of Y-Y direction was lain within the immediate occupancy level. It was observed that none of case, all the columns reached the slight structural damage state at third story.

6.LOSS ANALYSIS

Loss analysis was carried out only for structural elements and masonry walls because the cost of nonstructural elements and facility equipment are less compared to that of structural elements and masonry walls. Besides, simple calculation was done to determine losses rather than going for more rigorous analysis such as Montre Carlo type procedures because of time constraints.

The total structure repair cost of each case was calculated by multiplying the total repair quantity by the unit cost. The total repair quantity was estimated by identifying corresponding step of pushover analysis to story drift. It was observed that total structure repair cost would be between Rs.240,000.00 and Rs.2,500,000.00.The average total structure repair cost is Rs.2,337,550.00. However it was learnt that the total construction cost of this building is Rs. 16,000,000.00. Thus the average repair cost is 14.6% of total construction cost.

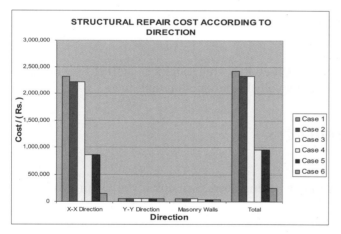

The variation of total structure repair cost of each case with direction of the structure was studied to identify which direction contributes most to the repair cost. The graphs of total repair cost versus direction and story are depicted in Figures 5.

Figure 5. Structure Repair Cost According to Direction

7.CONCLUSION

In summary, much knowledge has been gained from the study performed for this report. The behavior of three storied reinforced concrete frame structure used as a school building located in SriLanka, exposed to seismic load was studied.

It was found that the seismic performance of this structure is very poor in X-X direction while that of Y-Y direction is adequate. Therefore, even a small magnitude earthquake posed a threat because of poor X-X direction of the structure. The threat present with the occurrence of small magnitude earthquake events would most likely cause extensive damage to the structure.

As per FEMA 356, Prestandard and Commentary for the Seismic Rehabilitation of Building, performance levels of this building were defined using results of pushover analysis. The performance levels were indicated in the graph of base shear coefficient versus top drift. Accordingly, the inherent weakness of X-X direction of the structure was illustrated in the performance levels also. Therefore, these facts lead to a retrofit of the structure and it can be considered as a obligation for future work.

The behavior of masonry walls was studied using analytical model in structural analysis. It was found that the walls located at first two stories reached the complete structural damage state at small story drift. However, the behavior of the masonry walls was not studied for in case of out of plane failure. The loss analysis was carried out for masonry walls also but the repair cost of the masonry walls are negligible compare to that of reinforced concrete structure.

The repair cost of each damage state was calculated based on the rates obtained from a private company in SriLanka. Therefore, total structure repair cost was determined for maximum and mean cases of suite of earthquakes. It was found that total structure repair cost in X-X direction would be Rs.2,,228,000.00 in case of mean response story drift in the X-X direction while that of Y-Y direction is only Rs.56,000.00.

Considering the above facts, the following suggestions can be made as concluding remarks. Reinforced concrete shear walls could be provided to enhance the shear capacity in the X-X direction of the building. Besides, it was found that the base shear coefficient in Y-Y direction is 0.25 and it is adequate even for soft-first story structure. In addition, it is necessary to strengthen the infill masonry wall because it was observed that the infill masonry walls reached complete structural damage state prior to the structure reach immediate occupancy performance level.

8.ACKNOWLEDGEMENTS

My special gratitude to my advisor, Dr. Taiki Saito, Chief Research Engineer at IISEE, who has guided me from the inception of the course and given me valuable advices on this study.

9.REFERENCES

Abayakoon, S.B.S ,1998, Engineer, Journal of Institution of Engineers, SriLanka, Vol xxviii, No 2, 29-36.
Applied Technology Council, 2007, U.S. Department of Homeland Security, Federal Emergency Management Agency.
Federal Emergency Management Agency, 1998, Washington D.C.
Federal Emergency Management Agency, 2000, Washington D.C.
Federal Emergency Management Agency, 2003, Washington D.C.
Kabeysawa, T., Mostafei, H., Bulletin Earthquake Research Institute, University of Tokyo, Vol.79, 133-156.
The Japan Building Disaster Prevention Association.

Synopsis of Master Papers

Bulletin of IISEE, 43, 109-114, 2009

INVESTIGATION AND ESTIMATION OF STORY STIFFNESS AND DAMPING RATIO OF A 4-STORY WOODEN STRUCTURE BY MICROTREMOR AND MASS SLIDING SHAKER

Sithipat Palanandana*

MEE07157

Supervisor: Koichi Morita**

ABSTRACT

In this study, the structural dynamic properties of a 4-story wooden structure were investigated. Microtremor observation was carried out to determine the first and second natural frequencies of the structure. Based on these preliminary results, mass sliding shaker was mounted on the structure to apply harmonic force vibration in order to excite the resonant behavior of the structure. Target frequencies were applied to the mass sliding shaker to obtain the responses for all four modal shapes. Signal obtained from the measurement was then mathematically converted from time domain to frequency domain using Fast Fourier Transform for better interpretation to determine all four natural frequencies. Based on these results, resonance curves for all four modal shapes were constructed. Curve fitting based on mathematical formula was also drawn by changing the values of damping ratio and peak amplitude to fit the resonance curve. From microtremor observation data and time history responses of free vibration, damping ratio of the structure was determined for comparison. Based on curve fitting of resonance curves, natural frequencies and participation functions were also identified. Using story mass in addition to these identified properties, stiffness matrix could be determined. Finally, each story's stiffness was obtained by multiplying stiffness matrix by unit displacement vector.

Keywords: Microtremor, Fourier Spectrum, Resonant curve, Damping ratio, Story stiffness.

INTRODUCTION

For the past three decades, physical characteristics of structural systems have been focused on by many researchers and this is done through the testing procedure known as system identification. To be more precise, system identification can be described as mathematical tools and algorithm representing a dynamical model from measured data. This dynamical model is then used to identify the properties of structural systems in which researchers are interested. In the field of Earthquake Engineering, the properties of structural systems are essential to be identified. These properties are related to structural response subjected to the ground motion. In order to identify the physical characteristics of a structural system, microtremor observation has been widely used for decades. Microtremor observation generally provides clear amplification ratios for first and second natural frequencies. With ambient noises, the amplification ratios for higher mode are inaccurate or sometimes misinterpreted. The force vibration technique is then introduced especially for light structure such as wooden structure. Previous studies have shown that story stiffness of the structures are commonly obtained based on the values of first and second natural frequencies. For multi-degree of freedom structures, it is desirable to obtain natural frequencies for higher modes in order to give more accurate story stiffness. The purpose of this study is to estimate damping ratio and story stiffness of a 4-story wooden structure by using force

---

*Department of Public Works and Town & Country Planning, Thailand

** Senior Research Engineer, Building Research Institute, Tsukuba, Japan

vibration to stimulate the resonant behavior of the structure. Force vibration test also provides in depth modal analysis of the structure to obtain dynamic properties of the structure.

THEORY AND METHODOLOGY

System Configuration of the Building

The vibration of the building/structure during earthquake and microtremor is composed of several different types such as sway, rocking, relative displacement and several modes such as translations, torsion and those from higher modes. In general, the modes considered in vibration experiment are varied depending on type of building and the height of the building. In this study, the 4-story wooden building is investigated. Several high-sensitive accelerometers are placed at various positions inside the building. Each sensor is designated to catch the vibration signal along each story of the building as shown in Figure (1).

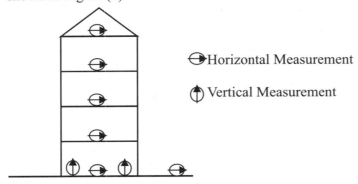

⊖Horizontal Measurement

⊕ Vertical Measurement

Figure 1. Sensors' location in wooden building

In general, the vibration of the building consists of rocking, sway and relative displacement. Each sensor was designated to measure the related motion. Rocking motion represents the different vertical displacement of the building. Sway motion is a relative displacement between ground and the building. Relative displacement represents the lateral deformation of each story with respect to another. For microtremor measurement, there are some assumptions needed to be made such as the floor is rigid, and ground surface is slightly separated from the building so that it is not affected by soil-structure interaction and so on.

Resonance Behavior of the Structure

The multi-degree of freedom structures possess several natural frequencies mostly corresponding to number of degrees of freedom. When the structure is subjected to the excitation, the resonance response is induced if the frequency of the excitation is close to each one of natural frequencies of the structure. In this study, a mass sliding shaker is used to implement the force function to the structure. The resonance test is carried out by changing the target shaking frequency using the oscillator generating the sinusoid function. Input and output responses are measured and data post-processing is implemented. The plots of ratios between input and output are fitted by the theoretical resonance curve. To illustrate this resonance behavior mathematically, the harmonic force at the forth story can be expressed as follows.

$$F = f \cos \omega t = (m_c a) \cos \omega t \tag{1}$$

in which m_c is the mass of the shaker and ω is frequency of input motion of the shaker. Output displacements are given by.

$$\{y\} = \sum_{s=1}^{4} {}_s u_4 \{{}_s u\} \frac{m_c a}{{}_s \omega^2 {}_s M} [A \cos {}_s \omega t - B \sin \omega t] \tag{2}$$

Therefore, the output acceleration is expressed as $\{\ddot{y}\} = -\omega^2 \{y\}$.

If only sth modal properties at ith story are extracted, response amplification in acceleration can be given as follows.

$$\left|\frac{\ddot{y}_i}{\ddot{x}_c}\right| = {}_su_{4\,s}u_i\left(\frac{\omega}{{}_s\omega}\right)^2\frac{m_c}{{}_sM}\sqrt{{}_sA^2+{}_sB^2} \approx \left(\frac{\omega}{{}_s\omega}\right)^2{}_s\beta_{\,s}u_i\frac{m_c}{m_4}\sqrt{{}_sA^2+{}_sB^2} \tag{3}$$

Therefore,

$$\left|\frac{\ddot{y}_i}{\ddot{x}_c}\right| = \left(\frac{\omega}{{}_s\omega}\right)^2{}_s\beta_{\,s}u_i\frac{m_c}{m_4}\frac{1}{\sqrt{\left\{1-\left(\omega/{}_s\omega\right)^2\right\}^2+4\,{}_sh^2\left(\omega/{}_s\omega\right)^2}} \tag{4}$$

where $\ddot{x}_c$, ${}_s\omega$, ${}_s\beta_{\,s}u_i$, and m_4 are input acceleration, natural frequency of the structure, participation function, and the 4th-story mass respectively.

Force Vibration Test Using Mass Sliding Shaker

In this study, mass sliding shaker or alternatively called Linear Shaker Seismic Simulation (LSSS) was used to induce the vibration on the structure. The installation of the shaker was done by fully attached the shaker at the middle of the roof floor using bolts and nuts. The reason is that the shaker can be considered as a part of the structure. All nine accelerometers were wired to the measuring device to measure the response at designated locations on the structure. One of nine sensors was attached on the shaker to capture the input motion given by the amplifier and function generator. The overview of the system configuration is shown in Figure 2.

The function generator (shown in Figure 2) generates the motion of the shaker at specified frequency and the function used was the sinusoidal function. The amplifier was used to increase the magnitude of the input signal by selecting appropriate bandwidth for better performance of the shaker.

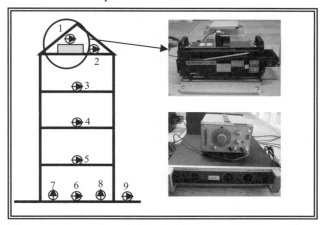

Figure 2. Overview of Shaker System Configuration

DATA ANALYSIS AND RESULTS

Microtremor observation was carried out initially to investigate the fundamental natural frequencies of the structure. The measurement was preferably done at midnight to minimize the unwanted ambient noise especially from the traffic. The measurement was recorded for 30 minutes to ensure the stable response of the structure. The sampling frequency was 200 Hz so the total number of data was 360,000. Once the raw data was converted to readable format, the response from each story can be displayed in time domain and frequency domain as shown in Figure 3 and 4.

From Figure 4 shown above, two obvious peaks of response can be observed. This implies that the first and second natural frequencies of the structure are most likely able to be spotted from the graph by projecting these two peaks on the x-axis. By doing so, these natural frequencies were found to be approximately 3.510 Hz and 11.730 Hz (dotted circles in Figure 4) for first and second natural frequencies respectively. These two frequencies were then used for the primary setup for the shaking frequency to determine the resonance frequencies. Range of target frequencies was applied to the shaker to identify all 4 natural frequencies. Fourier spectrum for each frequency was determined and

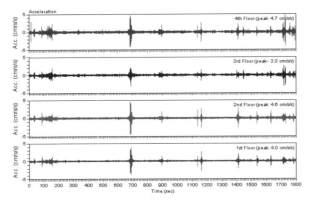

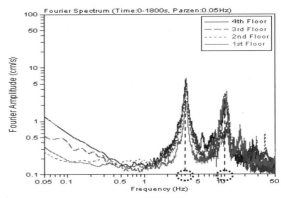

Figure 3. Time history response for each story   Figure 4. Fourier spectrum for each story

ratio between response of each story and shaker was plotted to form the resonance curve for each story. Then, by using Eq. (4), curve fitting can be done by selecting the values of damping ration and peak amplitude for the best fit to all resonance curves as shown in Figure 5.

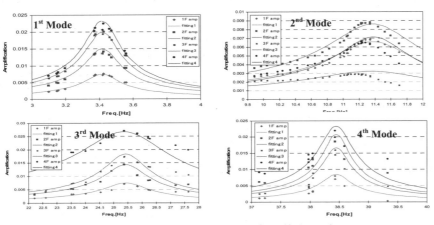

Figure 5. Resonance curves for all 4 modes

Based on the curve fitting method, damping ratio and all of the participation functions were obtained corresponding to each story for each mode. Phase lag of the response for each story was also determined to identify the modal shapes of the structure. The values of participation functions were used to plot the modal shapes as shown in Figure 6.

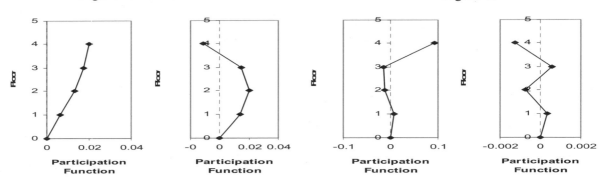

Figure 6. Participation Functions and Mode shapes

DAMPING RATIO AND STORY STIFFNESS

Half Power Method

In general, transfer function has a similar shape like a resonance curve. As described earlier, transfer function can be acquired from input and output data of microtremor observation. The frequency at the highest peak is natural frequency, f_0. .

The simplest method to estimate a damping ratio of a structure is called "Half Power Method".

In Figure 7, f_1 and f_2 can be acquired by projecting the value of $1/\sqrt{2}$ amplitude of the highest peak on the frequency axis. Then damping ratio can be estimated by

$$h = \frac{f_2 - f_1}{2f_0} = \frac{\omega_2 - \omega_1}{2\omega_0} = \frac{\Delta\omega}{2\omega_0} \tag{5}$$

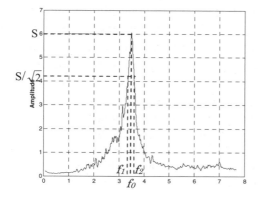

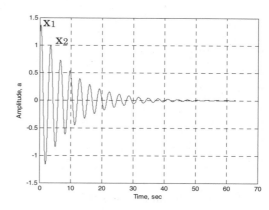

Figure 7. Transfer function   Figure 8. Free vibration

Random Decrement Technique

Curve fitting method and half power method are based on frequency domain, whereas time history domain methods are also used in several aspects. Considering two peak values (Figure 8) measured at two different positions with time interval T_d, the ratio of these two peaks x_1 and x_2 can be obtained as follows

$$\frac{x_2}{x_1} = \exp(-h\omega_0 T_d) = \exp(-h\omega_0 \frac{2\pi}{\omega_0\sqrt{1-h_2}}) = \exp(-h\frac{2\pi}{\sqrt{1-h_2}}) \tag{6}$$

Alternatively the damping ratio h can be approximately related to the logarithmic decrement as

$$h = \frac{\sqrt{1-h_2}}{2\pi}\ln\left(\frac{x_1}{x_2}\right) \approx \frac{1}{2\pi}\ln\left(\frac{x_1}{x_2}\right) \tag{7}$$

In most practical cases, several sets of logarithmic decrements are used for averaging. In general, time history domain may not be in the form of what is shown in Figure 8. The random signal is most likely presented for the structural response. In this method, the input signal can be formed by many sinusoidal waveforms. Then by superposing many part of output waveforms, only free vibration decrement can be acquired.

In this study, free vibration was obtained by inducing force vibration using the shaker and then the input motion is suddenly stopped providing free vibration afterward. This free vibration can be analyzed based on the concept of logarithm decrement. However, for the higher mode, free vibration response cannot be identified due to the ambient noise from the traffic. Therefore, in this section, 1st and 2nd mode free vibration response were considered to obtain damping ratios. Using all methods stated previously, average of damping ratios can be summarized as shown in Table 2.

Table 2. Damping ratios calculated by three methods

Story	Half Power Method		Random Decrement		Free Vibration	
	1st Mode	2nd Mode	1st Mode	2nd Mode	1st Mode	2nd Mode
Average	0.036	0.038	0.034	0.040	0.036	0.042

Estimation of Story Stiffness

By using the modal and spectral matrices, stiffness matrix can be given as follows:

$$[K] = \left([\Phi]_{norm} [\Omega^2]^{-1} [\Phi]_{norm}^{T} \right)^{-1} \tag{8}$$

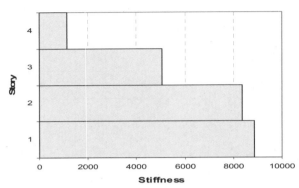

Figure 9. Story Stiffness

where $[\Phi]$ is the modal matrix consisting of transfer functions obtained from previous section and Ω^2 is spectral matrix consisting of diagonal values of eigenvalues, ω_n^2. To determine each story's stiffness, the concentrated equivalent story stiffness will be estimated by multiplying stiffness matrix [K] by the displacement vector whose elements are equal to 1. Each story's stiffness can then be obtained as shown in Figure 9.

CONCLUSIONS

Results from this study can be summarized as follows.

- Microtremor observation provides adequate information for fundamental natural frequencies of the wooden structure. The first and second natural frequencies are found to be 3.51 Hz and 11.73 Hz.
- The target frequencies are applied to the shaker to identify the resonance behavior of the structure. The resonant frequencies are found to be 3.42 Hz, 11.4 Hz, 25.45 Hz, and 38.45 Hz for first, second, third and forth modes respectively. It is noted that the resonant frequencies are slightly smaller than those identified by microtremor observation.
- Damping ratios are obtained based on several methods to compare the values. Results from these methods show that damping ratios do not vary corresponding to each mode.
- The forth-story stiffness calculated is quite small comparing with the others. The reason is that only shear deformation is considered in this study, whereas bending and torsional deformations are ignored.

Story stiffness and damping ratio can then be analyzed to provide the dynamic properties of the structure. These dynamic properties are used as the guideline to justify how sufficient the structure is and how the structure will perform during the earthquake.

ACKNOWLEDGEMENT

I would like to express my sincere gratitude to Dr. T. Kashima, my advisor, for his continuous support, valuable suggestion and guidance during my study.

REFERENCES

Hamamoto, T., 2003, Proceeding of SPIE, 5057, 106-117
Koyama, S., 2007, IISEE lecture note, Building Research Institute, Tsukuba, Japan
Churn, E. P., 2006, Master thesis, Dankook University,South Korea.
Morita, K., Teshigawara, M. and Hamamoto, T., 2005, Journal of Structural Control and Health Monitoring, 12, 357-380
Okawa, I., 2007, IISEE lecture note, Building Research Institute, Tsukuba, Japan
Niousha A., and Motosaka, M., 2007, AIJ Journal of Structural Engineering, 53, 297-304

Synopses of Master Papers   *Bulletin of IISEE, 43, 115-120, 2009*

QUALITY CONTROL OF RC ELEMENT FOR EARTHQUAKE DESIGN

Javier Yukio Yamamoto Munoz[1]   **Supervisor:**   Dr. Goto Tetsuro[2]
MEE07149   Dr. Taiki Saito[3]

ABSTRACT

Recently the RC structure has been increasing adopted in Dominican Republic; however, quality of the buildings in the rural area is poor. For this reasons my study is focused on the quality control of the structural elements to secure a better performance when an earthquake happen. By visiting several constructions site in Japan, the construction method, technology and the quality controls of their concrete work. By means of Comparing them with the methods used in the Dominican Republic and a proposal is prepared to improve the construction process and the quality of the structural elements in my country. Also studied how the honeycombs affect the concrete strength in the structural elements and a seismic capacity of high quality RC member. Based on the study result, methodology for better RC elements is proposed in this paper.

INTRODUCTION

A possibility is high that in a relatively near future the Dominican Republic is going to be affected by a strong seismic shock, since the last strong earthquake in 1946, 62 years has already passed. Despite the high seismic risk of our territory and despite the proximity of an important seismic event, the country is not prepared for an earthquake of considerable Magnitude. Since most of the population did not experience the last great earthquake of the August 4, 1946, the seismic risk in the Dominican Republic tends to be underestimated. This underestimation of the seismic risk allows construction of building in most inappropriate zones and quality of the buildings in the rural area is poorer. The informal construction of houses occupies 80% in all the country, this construction are done by the own owners or by neighbors of the place who do not count on any professional engineering formation. These types of buildings would be of the first to collapse at the time of an earthquake.

In the morning of September 22, 2003 the north zone of the Dominican Republic was hit by an earthquake of magnitude 6.5, the epicenter was located 15 kilometer from Puerto Plata city and 20 kilometer from Santiago de los Caballeros city. This earthquake damaged important buildings in Puerto Plata city, including schools and hospitals, and some buildings collapsed. After the earthquake, it was found that quality of these collapsed buildings were poor in terms of constructions materials and constructions works.

---

[1] Engineer, private company in Dominican Republic (**EPSA-LABCO S.A.**)
[2] Senior Researcher of National Institute for Land and Infrastructure Management (NILIM),
[3] Chief Research Engineer of Building Research Institute (BRI)

QUALITY OF RC CONSTRUCTION IN JAPAN

Constructions Practice

In the Dominican Republic concrete is most commonly utilized material for the constructions; the strength depends on several variables like characteristic of the materials, production method, transport, placement and cured. For these reasons part of this paper involves observing the construction process in Japan and exposing the problems of quality control that we have in the Dominican Republic.

Table 1 List of construction site visited in Japan

Location	Named	Detail
1	Mitakadai, TOKYO (UR Mitakadai complex)	8 story RC apartment building, Under Construction.
2	Oumigison, OKINAWA prefecture (Oumigison town office)	Reinforce Concrete office building, Constructed in 1925, under partial rehabilitation.
3	Miyakojima, OKINAWA prefecture (Irabu-oohashi)	RC bridge, Under Construction
4	Concrete Factory in Miyakojima	Automatic system to manufacture concrete severe quality control by computer monitoring

Important Factor for Construction Quality

Formwork System

The formwork is very important when we make out a structural element of reinforced concrete, since this form the required element. In both Japan and Dominican Republic wooden formwork is used.

The formwork system used in Japan is simpler and has better quality than in Dominican Republic. The wood used in Japan is processed and prepared to accomplish this function, which is more resistance and can be used more repeatedly. In order to join the formwork's woods in Japan steel separator is used to maintain the shape not to open while concrete is casted. In Dominican Republic use wooden lot is used but, is not resistant as the steel separator.

Manufacture of Concrete

Not all the concrete factories in Dominican Republic have high technology as the concrete factories in Japan, and the majority of these factories wit high technology concentrate in the capital or in major cities of Dominican Republic. Because in these places the most important projects are implemented and the volume of the consume concrete is more. In the rural area the concrete is made by the construction workers without any concrete design.

In Japan the concrete mixing is totally automatic, and the mixture is homogeneous wit all the production. In the Dominican Republic it is not guaranteed that concrete mixture, brought to the construction site is homogeneous. In many cases delivery concrete is delayed and for these reasons the driver of the truck adds additional water or additive to the concrete mixture before placing the concrete. For this reason it is necessary to make the concrete slump test to each concrete truck.

Qualified Staff

At Dominican Republic, the laborers do not have technical trainings and the majorities are Haitian immigrants, and in many instances they are not aware of what they are doing. For that reason they always go on with the same procedure of work and are very difficult to learn new techniques. To improve the quality in the construction site, capacitate the staff is important in order to increase quality and make a more fast work.

HONEYCOMB CONCRETE EFFECT TO THE CONCRETE STRENGTH

The specimens of concrete were given load and determine the Shear Stress to with and without honeycomb. The concrete strength is 21 N/mm², slump 18 cm and maximum coarse aggregate site is 20mm; the specimens were cast on July 2, 2008 and the actual test was carried out on August 5, 2008.

Shear Strength Test
Portable Structural Testing Equipment (PSTE)
Measuring instrument used is Data logger: TDS-303   b y Tokyo Sokki, Japan
Transducer: CDP50 by Tokyo Sokki, Japan.

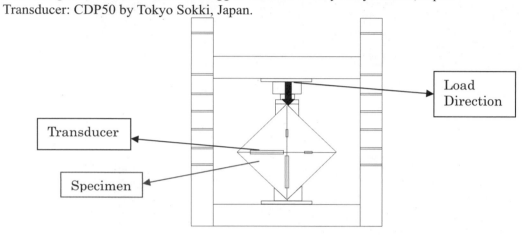

Figure 1 (PSTE) Developed by Dr. Goto Tetsuro of NILIM, Japan.

Specimens
The dimensions of the concrete specimen are 70cms x 70cms with 10cms of thickness. 2 type's specimens without honeycomb and with honeycomb concrete were prepared. The variable of this test is the honeycomb concrete area. Are tree type PH-1, PH-2 and PH-3.

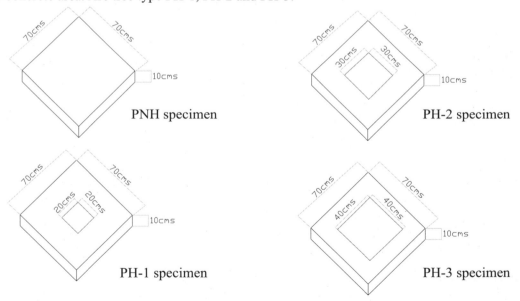

Figure 2 Specimen of Honeycomb

In order to create the honeycomb, sifter of 20mm is used to select large size aggregates. It results reduction of the quantity of mortar in the mixture.

Figure 3 Construction Process of specimens

Material test

Table 2 Quality control of concrete

Material	Cylinder Test		Specimens
	Compression	Tensile	Pulse Velocity
Concrete	24.62 N/mm²	2.15N/mm²	0.38 Km/sec

Compression Test
From the test results, the compressive strength and Young's Modulus of the concrete are listed in Table 2. The compressive strength is obtained by using the ratio of damage load fc and section area of the sample, and the Young's Modulus is calculated by using the secant value of $0.5*fc$. The compressive strength and Young's Modulus are about 25MPa and 40kMPa respectively as shown in Table 1

Splitting Test
The splitting test is about 2.15 N/mm² respectively as shown in Table 2.

$$\sigma t = \frac{2P}{\pi dl}$$

σt: Tensile strength
P: Maximum load
d: Diameter of cylinder
l: Length of cylinder

PUDINT Test
In order to check the homogeneity of the concrete in non honeycomb concrete area, an Ultrasonic non-destructive integrity testing (PUDINT) are used. This test is applied to all Concrete plate but not in the concrete honeycomb concrete area.

As velocities, showing in the Table 2, do not have drastic change in the tested areas. That means the concrete preserve the uniformity in all the concrete area except for the honeycomb area.

Shear Test Results
Figure 4 shows the relationship between vertical deformation and load obtained from shear stress test for 4 specimens.

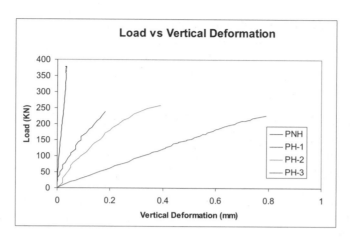

Figure 4

Table 3 result of shear stress test

Specimen	Load (KN)	Qu (KN)	Area (mm²)	Shear Strength (N/mm²)	Vertical Deformation (mm)
PNH	379.5	268.35		3.83	0.03
PH-1	238.37	168.55	70000	2.41	0.18
PH-2	259	183.14		2.62	0.39
PH-3	226.75	160.34		2.29	0.79

From figure 4 it is observed that vertical deformation of specimen with honeycomb become large compared with specimen without honeycomb. The specimen whit large honeycomb area, PH-3, gives the largest vertical deformation.

Figure 5 shows the graph of shear strength of the specimens.

From this graph, the ratios of shear force reduction are 1.43 for PH-1, 1.21 for PH-2 and 1.54 for PH-3 in average, the honeycomb reduce 36.4%.

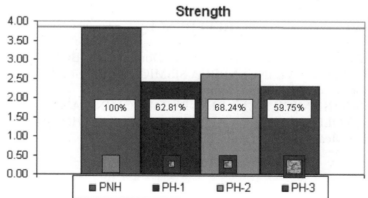

Figure 5

When comparing the results between them with honeycomb specimen and without honeycomb specimen. The load capacity of PNH is higher than the PH-1, PH2 and PH-3. The Vertical Deformation of the PH-1, PH2 and PH-3 specimen is proportional to the honeycomb area.

The honeycomb affected the concrete the shear strength of de concrete and de vertical deformation of the specimens

CONCLUSION

From this study, it is conclude the honeycombs in the structural element affect the strength of the concrete. The honeycomb area is an important factor in this aspect. However, it is ideal to test specimens of structural elements with steel bars as adopted in the actual structural body. Therefore, the challenge for the future would be testing structural elements with the same honeycomb to examine the effects of honeycomb to the structural elements. Also step from this study is to verify the repair methods used for honeycomb, and conventional repair methods should be tested to verify their effectiveness. In this study specimens used are not structural elements. When comparing the construction methods of the Dominican Republic with those of Japan, the problems at the construction site in the Dominican Republic are clearly observed as follows.

o   Absence of quality control
o   Ignorance of regulation for quality control.
o   Expansion of buildings often made ignoring original structural design.
o   Absence of qualified staff.

These problems can be minimized with proposals of quality control on the construction site. The proposal includes training and education to the construction workers in term of quality control.

Following recommendations are presented:

o   Educate the workers, designers and constructors in fulfilling the standards established by the construction code.
o   The government has to introduce programs capacitating the workers.
o   Investing to new technologies, which help to secure the concrete quality such as assembling the formwork.

ACKNOWLEDGEMENT

I would like to express my sincere gratitude to Organization of Urban Renaissance Agency (UR), The Morimoto Construction Company, the Oomigi Town Office of Okinawa prefecture government, Miyakojima Branch Office of Okinawa prefecture government, National Registered Architect Association of Miyakojima Branch of Okinawa prefecture and Daiyomi Construction Company.

REFERENCE

Ing. Héctor E. O'Reilly Pérez, Reglamento Sísmico Dominicano, BOLETIN DE LA SOCIEDAD DOMINICANA DE SISMOLOGIA E INGENIERIA SISMICA (SODOSISMICA)
Ramón A. Delanoy, Aspectos de Sismología Dominicana, 1995
R. Osiris de León, RIESGO SÍSMICO EN LA REPUBLICA DOMINICANA
Tadashi Maki, SENIOR VOLUNTEER, JICA ADVISER (ABTI-EARTHQUEAKE) ,SEISMICITY in the Dominican Republic- For Measurement of Anti- Earthquake-2004.

Synopsis of Master Papers   *Bulletin of IISEE, 43, 121-126, 2009*

STUDY ON TSUNAMI DATABASE FOR TSUNAMI EARLY WARNING SYSTEM IN BANGLADESH

A.K.M.Ruhul KUDDUS*   **Supervisor: Yushiro FUJII****
MEE07173

ABSTRACT

In this study, numerical tsunami simulations along the Bangladesh coast were conducted in order to make tsunami database. The numerical model is based on the spatial grid system in Cartesian coordinate system using a non-linear tsunami theory. Two types of bathymetry grid interval were considered in this study such as (1) one arc minute, that is about 1850 m and (2) twenty arc seconds that is about 616.67 m. The source points have been set up near the Bangladesh coast and parallel to the trench. The coastal points, forecast points and one tide gauge station were set up near the coast. Ten source points were selected with four different magnitudes (M_w 6.5, 7.0, 7.5 and 8.0) and four different depths (0, 10, 20 and 30 km). Tsunami heights were checked graphically for all cases. The Green's Law was applied to obtain the reliable tsunami heights along the coast from the tsunami heights at forecast points. Results of both grid interval of bathymetry were compared with each other. The simulation results were stored in MySQL system for a tsunami database.

Keywords: Tsunami arrival time, Tsunami height, Tsunami database.

INTRODUCTION

Bangladesh is surrounded by the regions of high seismicity, which include the Himalayan Arc and Shillong plateau in the north, the Burmese Arc, Arakan Yoma anticlinorium in the east, and complex Naga-Disang-Jaflong thrust zones in the northeast (Hossain, 1989). The northeast side of the Indo-Australian plate forms a subducting boundary with the Eurasian plate on the border of the Indian Ocean from Bangladesh. The long seismic gap from Bangladesh coast to Andaman Islands of the bay of Bengal could produce a big tsunami in future. Southern part of Bangladesh is a coastal region. Millions of people are living in this coastal region. Considering the potential seismic source (Cummins, 2007), a tsunami in this region poses a very significant threat to a large coastal population. This study has been addressed a tsunami database for tsunami early warning system, which will prevent human lives and reduce damage from the tsunami.

METHOD OF ANALYSIS

Bathymetry Data

The bathymetry data were downloaded from the web site of GEBCO. Two types of grid interval of bathymetry data, one arc minute that is about 1850 m and twenty arc seconds that is about 616.67 m, were used in this study. GEBCO data have positive values on land grids and its grid format is NetCDF. It was needed to convert another format by using some GMT commands. The converted bathymetry grid data became negative values for land (-999) and positive values for sea depth.

---

*Bangladesh Meteorological Department, Bangladesh.
**International Institute of Seismology and Earthquake Engineering (IISEE), Building Research Institute (BRI), Japan.

Magnitude and Depth

The four magnitudes and four depths have been set for the computation of tsunami simulation for each source point in order to make database. The magnitudes are M_w 6.5, 7.0, 7.5 and 8.0 with the interval of 0.5. The depths are 0, 10, 20 and 30 km with the interval of 10 km.

Output Points and Source Points

Twenty-two coastal points, twenty-two forecast points and one tide gauge station were considered as the output points that were set up near the coast. A minimum depth was fixed to 1 m for the coastal points and one tide gauge station. The forecast points were fixed at 10-minute ≈ 18.5 km away from the coast. The output points are shown in Figure 1 with contours of depth.

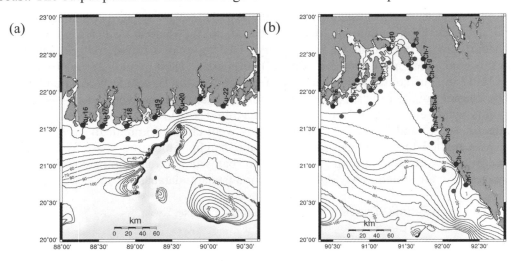

Figure 1. Bathymetry depth contour map near Bangladesh coast. Red circles and Maroon circles indicate forecast points and coastal points, respectively.

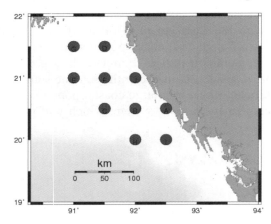

The source points were set up near and parallel to the trench. It was fixed according to grid size (thirty-minute interval). Red circles in Figure 2 represent source points. The total number of source points is ten.

Figure 2. Location of source points.

Fault Parameter and Seafloor Deformation

Fault parameters, length, width, and displacement were calculated by using scaling law (Tatehata, 1997) as a function of magnitude. The strike of 320° is set, which is parallel to the trench, considering that most of the earthquakes happen along the trench; the dip angle is 45°, slip angle is 90° and number of fault segment is one. The dip and slip angle are set at these values because it is the worst-case scenario of tsunami. The location of the hypocenter is assumed to be in the middle of the fault plane.

The depth of top left corner (TLC) of each fault was used as the input depth in order to compute tsunami simulation.

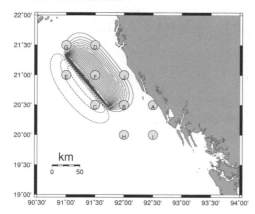

The seafloor deformation due to fault was calculated by Okada's formulas (Okada, 1985), using the fault parameters. The vertical deformation of the sea bottom was used as the initial condition for the numerical tsunami simulation. The initial size of tsunami depends on the amount of vertical seafloor displacement. It is determined by the magnitude of earthquake, depth and fault plane mechanism. Figure 3 represents deformation area of source point F for magnitude of 8.0. The red contours denote uplift with the contour interval of 0.1 m, while the blue contours denote subsidence with the contour interval of 0.1 m.

Figure 3. Vertical displacement of the seafloor.

Numerical Tsunami Simulation

TUNAMI-N2 code is applicable in order to compute a tsunami propagation. The simulation program (e.g. Fujii, 2008a) is carried out considering the bottom friction and using the non-linear term. The area of computation is determined by considering the source points and coastal points. The area is 15° N to 23°N and 85°E to 95°E. The number of grid points is 601 and 481 in x and y direction, respectively. The calculation time is set to12 hours for the area. The time step is set to 3 s. Number of time steps for snapshots is 200 (10 min). Tsunami heights at the coast are calculated by using Green's law from tsunami heights at the forecast points.

Inverse Refraction Diagram

Refraction diagrams were used to calculate the tsunami travel times by using TTT (e.g. Fujii, 2008b). TTT originally developed by Paul Wessel, Geoware and modified by UNESCO (TTT Software Package).

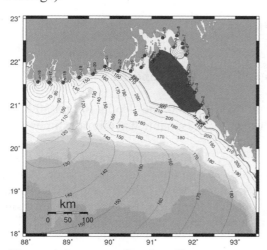

Refraction diagram can be drawn backwards from the coast. Such a diagram is called inverse refraction diagram and is sometimes used to estimate the tsunami source area. It is prepared to calculate inverse travel times of tsunami from the coastal points to the source. The grid points that have absolute values greater than 0.04 m are considered as the source points inside the seafloor deformation area. The red color of Figure 4 indicates the grid points. Light black curves are the travel time arcs computed for three-output points. Red area indicates assumed tsunami source. The contour interval is ten minutes. The difference between any two contour lines of inverse refraction diagram indicates that tsunami wave velocity is dependent on water depth.

Figure 4. Inverse refraction diagram.

RESULT AND DISCUSSION

The results of tsunami simulation (tsunami heights and tsunami travel times at the output points) are stored first into Excel sheets in order to make a database.

Tsunami Heights

The simulation results are represented in Figure 5 for checking. One is tsunami height versus magnitude with constant depth and another one is tsunami height versus depth with fixed magnitude.

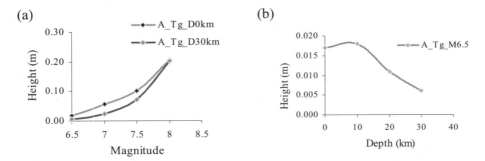

Figure 5. (a) Magnitude versus tsunami height with constant depth.
(b) Depth of fault center versus tsunami height with fixed magnitude.
A: source point, Tg: tide gauge station, D: depth, and M: magnitude.

Figure 5(a) shows that as the magnitude increases the tsunami height also increases. Figure 5(b) shows that as the depth increases the tsunami height decreases. Simulation results are checked for all cases and all coastal points. Then we found that big magnitude together with shallow earthquake produces large tsunami.

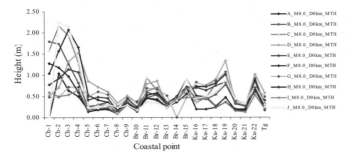

According to the simulation results, tsunami amplitudes become the maximum at the coastal points that are closer to the source and tsunami amplitudes become the minimum at the coastal points, which are far from the source. Depending upon magnitude, depth and location of source, Figure 6 shows that the tsunami heights along the coast are different.

Figure 6. Tsunami heights against the coastal points for different tsunami sources with the same magnitude and depth.

Comparison between Tsunami Travel Times from Tsunami Waveforms and TTT

Depending upon different source points with the same magnitude M_w 8.0 and depths (0 and 30 km), Figure 7 shows the comparison between tsunami travel times from tsunami waveform's times and tsunami travel times calculated by using TTT software.

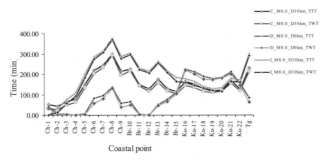

The tsunami travel times are almost similar for the same case. Tsunami travel times are difficult to detect from the first small change of waveforms at the coastal points.

Figure 7. Comparison between tsunami travel times using waveforms and TTT against 22 coastal points and one tide gauge station.

Tsunami Heights with Green's Law

GEBCO, one arc minute data, is not good enough to calculate accurate tsunami heights along the coast. The Green's Law, energy conservation law along the ray (Satake, 2008), is needed to obtain the reliable tsunami heights along the coast from the forecast points at the sea. Tsunami heights at the coast (H_1) are calculated by tsunami amplitudes at the forecast points (H) multiplied by the fourth root of the sea depth ratios at the forecast points (h) and at the coastal points (h_1). The sea depth at the coast (h_1) is assumed 1 m. Hence, tsunami heights along the coast becomes,

$$H_1 = \sqrt[4]{h}H \qquad (1)$$

Depending upon different source points with the same magnitude and depth, the tsunami heights calculated by applying Green's Law are shown in Figure 8.

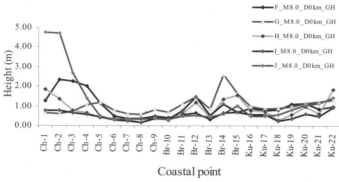

The maximum tsunami height with Green's Law against the coastal point (ch-1) is estimated 4.8 m. This coastal point shows the highest tsunami height with Green's Law as it is closest to the tsunami source.

Figure 8. Tsunami heights with Green's Law for different tsunami sources with the same magnitude and depth.

Finer Grid Computation

Twenty arc second grid intervals (~ 616.67 m) of bathymetry data are applicable to compute the tsunami simulation for the same area. The twenty arc second grid intervals of bathymetry data are re-sampled from GEBCO one arc minute grid data. The number of grid points is 1801 and 1441 in x and y direction, respectively. The time step is set to 2 s. Number of time steps for snapshots is 300 (10 min). Figure 9 shows that tsunami heights for the twenty arc second grid interval are little higher than those for the one arc minute grid interval, because the grid size is smaller. Finer grid computation can express shorter wavelength of tsunami than coarser one. The black curve denotes tsunami heights against coastal points for the twenty arc second grid interval; the green curve denotes tsunami heights against coastal points for the one arc minute grid interval.

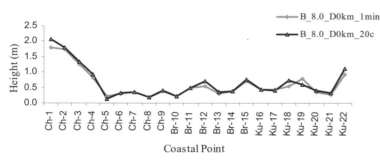

The simulation time required for the twenty arc second grid interval is four hours for each case. Meanwhile the simulation time required for the one arc minute grid interval is twenty-one minutes for each case. The computation with the one arc minute grid interval of bathymetry data and Green's Law were useful to save time of about 474 hours.

Figure 9. Comparison of tsunami heights for twenty arc second grid interval and one arc minute grid interval.

DATABASE

MySQL Database

Simulations data is put into a MySQL database by using some MySQL commands. The simulation result is divided into two tables under the name of "BMD_tsunami" main table, that is, "Hyp_Ruhul" table and "Simudata". "Hyp_Ruhul" table contains all information regarding hypocenter parameters as latitude (°), longitude (°), depth (km), magnitude, slip (cm), length (km), width (km), strike (°), dip (°), rake (°) and vertical displacement (m). Output points together with simulation results are stored at the "Simudata" table. "Simudata" table contains the number of output points, name of output points, longitude (°), latitude (°), depth (m), tsunami travel time (min), travel time of maximum tsunami heights, maximum tsunami heights (m), Green's heights (m), FILE_NAME.

The method of database system used in this study is important in order to issue tsunami warning. The numerical simulation takes a long time, if we ran the simulation after the occurrence of an earthquake, tsunami would arrive at the coasts before tsunami warning is announced.

CONCLUSION

In this study, a prototype tsunami database in Bangladesh is created. Simulation results are stored into the MySQL database. When an earthquake occurs, the tsunami heights and tsunami travel times along the coastal area will be obtained by interpolation of the data using the location of epicenter, magnitude and depth from the database for the information of tsunami warning.

According to the analysis of simulation results, it is found that big magnitude together with shallow earthquake produces large tsunami. The coastal points, which are more far in distance from the source, have the smallest tsunami heights and take the longest tsunami travel times. On the other hand, the tsunami sources nearest to the coastal area cause higher tsunami heights. Tsunami heights computed for the finer grids (20 arc second grid) are slightly higher than that for the one arc minute grids against the same coastal points because the grid size is smaller. Finer grid computation can express shorter wavelength of tsunami than coarser one.

ACKNOWLEDGEMENT

I would like to articulate my heartfelt gratitude advisor Dr. B. Shibazaki, for his heartiest and spontaneous support, valuable comments and guidance during my study.

REFERENCES

Cummins, Phil R., 2007, Nature 449, 75-78 doi: 10.1038/nature06088.
Fujii, Y., 2008a Tsunami Simulation, 2007-2008 IISEE Lecture Note, BRI.
Fujii, Y., 2008b Tsunami Source, 2007-2008 IISEE Lecture Note, BRI.
Hossain, KM., 1989, 7th Geological Conference, Bangladesh Geological Society.
Okada, Y., 1985, Bulletin Seismological Society of America, 75, 1135-1154.
Satake, K., 2008, Tsunami Generation and Propagation, 2007-2008 IISEE Lecture Note, BRI.
Tatehata, H., 1997, The New Tsunami Warning System of The Japan Meteorological Agency, JMA, Japan.

STUDY ON TSUNAMI NUMERICAL MODELING FOR MAKING TSUNAMI HAZARD MAPS IN INDONESIA

Athanasius Cipta*   **Supervisor: Fumihiko IMAMURA****
MEE07172

ABSTRACT

Five source models along west coast of Sumatra and one for the south coast of Java were conducted for tsunami simulations. Based on those results, waveform characteristics were analyzed. We chose the west coast of Sumatra for waveform analysis because the coast has variation in water depth, morphology and shoreline shape.

For making tsunami hazard map, we chose source parameters from the 2006 south Java tsunami event. The target area is Pangandaran Peninsula, one of the most damaged areas attacked by the 2006 event. We will compare the simulation results with the field survey results by Kongko et al. (2006). We want to know how appropriate the simulation results are for the tsunami earthquake event.

To simulate near field tsunami propagation, numerical modeling was used. We considered nested areas, in limited area around Padang (Sumatra cases) and Pangandaran (Java case). From simulations, we got results such as tsunami waveforms, run-ups, tsunami heights and inundations. Based on these results, field surveys, and some pictures taken from AMS (American Marine Survey) and Google Earth, we made tsunami hazard maps.

Keywords: Tsunami, Simulation, Inundation, Hazard Map.

INTRODUCTION

The northern Sumatra and the south coast of Java Island were hit by tsunamis which led to wave run-ups and landward inundations. The devastation at any particular location is caused by a function of the velocity, acceleration, and elevation of the water as it interacts with natural and man-made coastal objects. Clear understanding of tsunami wave behavior is indispensable to tsunami hazard assessment. Decisions affecting human safety require systematic methods for evaluating the tsunami events.

DATA AND METHOD OF COMPUTATION

Bathymetry and Topography Data

Bathymetric and topographic spatial data are the basic data for tsunami simulation, as we know tsunami wave propagates over the sea bathymetry and when tsunami wave inundates inland, floods over land topography.

For Sumatra cases, we used GEBCO (GEneral Bathymetric Chart of the Ocean) 1 arc min bathymetry for region1, nautical chart 20 arc seconds for region 2 and nautical chart 5 arc seconds for region 3. SRTM (Shuttle Radar Topographic Mission) is also used for topographic data in region 3.

For Java case, GEBCO is used for both of the bathymetric and topographic data. We used 1 arc min grid size of bathymetry and topography for region 1 and grid size of 15 arc seconds for region 2.

---

*Volcanological Survey of Indonesia
** Professor, Disaster Control Research Center, Tohoku University, Japan.

Tsunami Simulation

Tsunami simulation is recognized to be an essential tool to explain the observations and records of a tsunami (tsunami heights, travel times etc.), and to assess tsunami hazard, vulnerability and risk. Tsunami simulation can be used to provide tsunami assessment and prediction of arrival times, expected wave amplitude and coastal effects.

For these purposes we use tsunami simulation program named TUNAMI code, developed by Tohoku University (Imamura et al., 2006; Koshimura, 2008) and the Boussinesq approximation model. Cartesian coordinate system will be used in numerical simulation and the shallow water theory with bottom friction in the near-shore region in which water depth is shallower than 50 m (Nagano et al. 1991).

The most popular stability criterion is Courant-Friedrich-Levy number (C.F.L. condition) which states that the time step must be smaller than the time it takes for a wave to propagate from one grid point to the next. Spatial and temporal grid sizes are set to satisfy this stability condition in the numerical computation to avoid instability results.

The fault size parameter such as length, width, and slip amount (dislocation) can be determined by using scaling law theory for dip-slip fault in subduction region (Papazachos et al., 2004). Scaling law is useful for calculating fault parameters which is controlled by magnitude (Mw).

For Sumatra cases, we assumed four sources at 102.1°E and 6.39°S (fault 0), 100.25°E and 4.25°S (fault 1), 98.48°E and 2.14°S (fault 2), 96.71°E and 0.03°S (fault 3). All the cases have parameters: Mw 9.0, fault length 575.44 km, fault width 144.54 km, slip 2.5 m, top depth of fault 10 km, strike 320°, dip 10° and rake 12°. Fault 4 is assumed at 100.0°E and 1.0°S with the fault parameters: Mw 8.5, fault length 300 km, fault width 79 km, slip 6.0 m, top depth of fault 10 km, strike 320°, dip 10° and rake 12°, although the fault 4 is not a realistic case and only for simulation. Tsunami propagation was calculated for 24 assumed tide gauge (TG) stations as outpoints along the west coast of Sumatra. Tsunami sources from fault 2 and fault 3 are used because of the biggest effects to Padang City (target area) and Mentawai Islands, off Padang shoreline.

For Java case, we assumed an earthquake source at 107.82°E and 10.285°S, Mw 7.7, fault length 80.9 km, fault width 40 km, slip 2.5 m, depth of fault 20 km, strike 289°, dip 10° and rake 95° (Yukselme, 2006). Tsunami propagation was calculated for 13 assumed TG stations, along the affected area by the 2006 South Java tsunami.

RESULT AND DISCUSSION

Sumatra Cases

Tsunami Wave and Water Depth
Figure 1 shows the location of TGs and assumed faults. Each TG is located at the different depth that will affect to calculated tsunami waveform.

In the shallow sea, amplitude of tsunami wave is getting higher, wavelength and celerity are decreasing. Figure 2 right (shallow water) shows that amplitude is higher than in Figure 2 left (deep sea). These oscillations are the result of reflection. Padang TG is located inside the channel, between mainland and islands arc west of Sumatra, tsunami wave that goes inside the channel approaches mainland and is trapped inside the channel. Direct wave resonances with reflected wave, second reflected and so on. In this case shoreline acts as wave reflector.

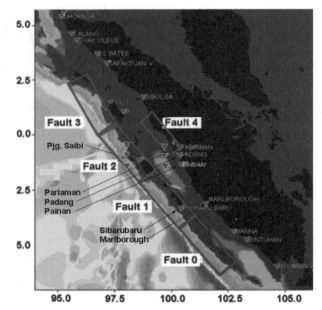

Figure 1. Location of the assumed faults.

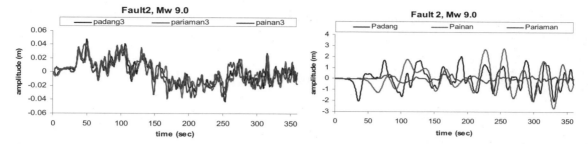

Figure 2. Tsunami waveforms affected by water depths, calculated at Padang, Painan and Pariaman TGs.

Tsunami Wave and Seafloor Morphology

Figure 3 shows different tsunami waveforms, affected by seafloor variation. Sibarubaru TG is facing to the open sea and located at shallow water. This conditions cause tsunami wave has higher amplitude and shorter period. The first incoming tsunami wave was not the maximum tsunami, the maximum one came later. The amplitude and period of tsunami wave changed after 225 min and amplitude became lower and period became shorter. After 275 min, amplitude and period increased and tsunami wave reached the maximum in height. The complexity of seafloor morphology causes oscillation wave, amplitude change and period change.

Figure 3 Tsunami waveforms calculated at Sibarubaru Island, Marlborough and Pjg. Saibi.

Sibarubaru is located at the southernmost of this passage, facing directly to the open sea and Marlborough is located at the southernmost of the passage (Figure 4). Tsunami wave hits Sibarubaru and rapidly attenuates. After 150 min, amplitude of tsunami wave becomes lower and lower. Marlborough TG shows that wave periods changes by time, from 100 to 150 min period is larger than before. After 150 min, period becomes smaller and increases again after 250 min. First tsunami wave went inside

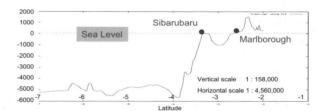

Figure 4 Cross section Sibarubaru-Marlborough, passage between Sibarubaru and Marlborough.

passage, hit Sumatra Island, reflected and resonance with next incoming tsunami wave. Amplitude of tsunami wave does not attenuate rapidly because tsunami wave is trapped inside the passage.

In front of Pjg. Saibi TG, there are 2 sea ridges, both on the right and left side, form a passage, hereupon, tsunami wave is concentrated in this passage and propagate to Pjg. Saibi. After 150 min, tsunami wave amplitude tends to decrease and becomes lower gradually.

Tsunami Wave and Shoreline Shape

Maximum tsunami height varies depending on the location of each TG. Different location gives different result as shown in Figure 5. In Padang, tsunami wave amplitude was slightly lower than in Pariaman and Painan because Padang TG is located at the straight coast line, although Pariaman is located at straight coastline but at the shallower water than Padang TG.

Maximum tsunami heights in Pariaman and Painan do not show significant difference even though these cities are located in different shoreline shapes. Pariaman is located in straight shoreline and Painan is located inside the shallow circular bay. Even Painan is located inside the bay, but tsunami wave did not concentrate in this bay because this bay is too wide and beach slope is too steep.

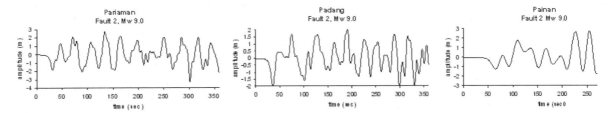

Figure 5. Tsunami waveforms calculated in Pariaman, Padang and Painan TGs.

Tsunami Wave and Location of Fault

Tsunami wave calculated in Afulu TG has very high amplitude for fault 3 case. One of the reasons is that the position of earthquake source is located parallel to Afulu, and tsunami wave propagated from south to the north, then many times tsunami waves reflected on beach of Sumatra, reflected wave resonance with later incoming wave and produced higher amplitude wave. Figure 6 shows that maximum tsunami height came later and tsunami wave did not attenuate rapidly.

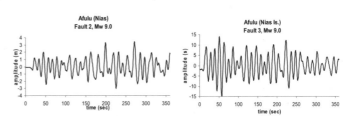

Figure 6. Tsunami waveform recorded in Afulu TG.

Tsunami wave has short period because it propagated in shallow water. Tsunami wave did not attenuate rapidly because the reflected wave came, reflected and came again and sometimes made resonance with later incoming wave

Java Case

Tsunami Travel Time

Tsunami travel time (Figure 7) is time between earthquake occurrence time and tsunami wave arrival time. The shorter the tsunami arrival time, the nearer the coastal area, many evacuation facilities should be built. Tsunami travel time is important also for people to decide to evacuate themselves as soon as possible.

Tsunami wave hit Pangandaran about 30 min after the earthquake occurred. If tsunami early warning was issued 5 min after the earthquake, it would mean that people near coastal area have only 25 min for evacuation.

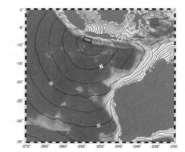

Figure 7. Tsunami travel time.

Inundation Area

Naturally, inundation area breadth is controlled by tsunami height, force of tsunami, inland morphology, and stream pattern. Variation of feature of earth surface gives various possibilities for inundation.

The coastal areas are lying below the mean waterline due to its down sloping characteristics. Furthermore, the back water system is generally running parallel to the shoreline and the coastal areas are like a narrow lane of land (barrier beach) lying between the backwater and sea. This condition caused tsunami inundated farer and wider inland because tsunami wave inundated through streamline.

Maximum Inundation Height

Maximum inundation height is controlled by shoreline shape and beach slope. Figure 9 and 10 show maximum inundation heights. Maximum inundation height got up to 1.32 m in Nusa Kambangan that is located perpendicular to the fault length.

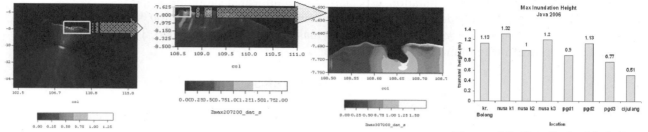

Figure 9. Maximum inundation heights.

Figure 10. Inundation heights at several villages.

Tsunami Hazard Map

Tsunami hazard map contains information about damages of tsunami, affected area, emergency information, other information and graph, depend on purpose.

In tsunami hazard map, we put some important area that may be inundated, like business center, airport and area with special feature that could be flooded by tsunami wave higher than other area. We also put tsunami waveform and tsunami height in some TGs. Based on simulation data such as, waveform, travel time, inundation height, and another data, such as survey data, photograph, map image, we can make tsunami hazard map (Figure 11).

Some pictures in Pangandaran tsunami hazard map (Figure 11) show interesting places that were attacked by the 2006 tsunami. Pangandaran coast, the most valuable tourism destination in West Java Province is the interesting place that was attacked by tsunamis twice in last hundred years (Tedy Eka Putra, LIPI, personal communication). Based on field survey (Kongko et al., 2006), 2 tsunami deposit layers are found in Cikembulan Village, Pangandaran.

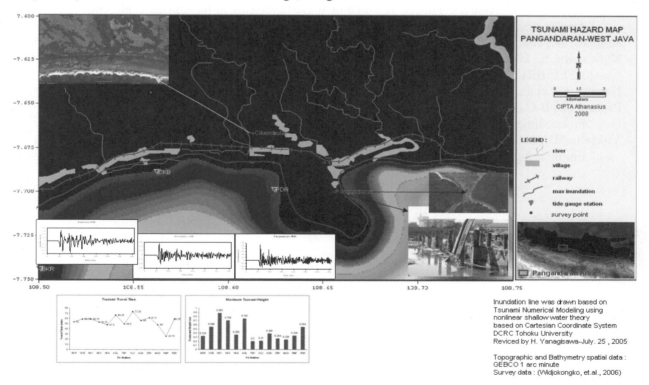

Figure 11. Example of tsunami hazard map, Pangandaran area, West Java Province.

CONCLUSION

In this study non-linear theory model based on nested grid system computations of tsunami propagation is conducted. The precise and finer (< 3 arc seconds) bathymetry and topography data are important as basic input for simulation, especially for complicated shoreline shape.

Tsunami height at tide gauge stations, tsunami travel time, inundation height and inundation area are some of many factors to be considered in making tsunami hazard map. Tsunami hazard map is the first step in effort to minimize tsunami disaster.

FUTURE WORKS

There are still broad possibilities to improve tsunami simulation based on the nested grid system technique in this paper. The important things to be done in the future are:

- Improvement of computed results for tsunami travel times, tsunami heights and inundation heights in the model, the detailed and accurate bathymetry data is essential for tsunami numerical computation.
- Improvement of modeling by using more accurate and effective input source parameters.
- Topography data should be concerned in the numerical computation where interaction with the structure as well as vegetation or buildings should be taken into consideration in calculation of tsunami coastal effect.

ACKNOWLEDGEMENT

We would like to express our sincere gratitude to advisor Dr. Y. Fujii of IISEE, BRI for continuous support, valuable suggestion and guidance during study.

REFERENCES

Imamura, F., et al., 2006, Tsunami Modelling Manual (TUNAMI model).

Mofjeld H.O., 2000, Analytic Theory of Tsunami Wave Scattering in the Open Ocean with Application to the North Pacific, NOAA Technical Memorandum OAR PMEL-116.

Kongko, W., et al., 2006, Rapid Survey On Tsunami Jawa 17 July 2006, BPDP-BPPT & ITS.

Koshimura, S., 2008, Lecture Note on Theory of Tsunami Propagation and Inundation Simulation, IISEE/BRI.

Nagano, O., et al., 1991, Natural Hazards 4, 235-255.

Papazachos B.C., et al., 2004, Bulletin of the Geological Society of Greece vol. XXXVI, 2004, Proceedings of the 10th International Congress, Thessaloniki.

Yukselme D., 2006, July 17, 2006 Indonesia Java Tsunami, Middle East Technical University, Civil Engineering Department, Ocean Engineering Research Center, Ankara.

fA PROTOTYPE OF WEB-APPLICATION FOR TSUNAMI DATABASE ALONG SOUTHERN JAVA ISLAND COASTLINE

Ariska Rudyanto[*]
MEE07170

Supervisor: Yohei HASEGAWA[**]
Yosuke IGARASHI[**]
Yushiro FUJII[***]

ABSTRACT

Development of tsunami database system along coastline area of southern Sumatra, Java and Bali islands was presented in this study. This was performed in three stages, there are: (1) applying tsunami numerical simulation; (2) constructing a database; and (3) retrieving tsunami heights and arrival times from the database for assembling tsunami warning message. First stage was conducted by employing TUNAMI-N2 and TTT software for 1 real case and 28 artificial cases (with 4 depths of 0, 20, 40 and 60 km; and 3 different magnitudes of 7.0, 7.5, and 8.0), to obtain tsunami heights and arrival times on 56 coastal points. The results of tsunami heights at the coasts were considered applying Green's law at 50 m sea depth of forecast points. Then, a web application system using Apache web server, MySQL relational database and PHP programming language, were used to perform the second and third stages. The tsunami database which was successfully constructed in this study consists of 338 tables, and is divided into 3 parts: coastal point table, hypocenter table and simulation data table. Relation between the tables in the database based on the primary key from each table is a '*one to many*' type. The system retrieves tsunami heights and arrival times from the database and assembles warning messages. For retrieving methods, simple and interpolation techniques, are used. Lastly, it was found that the web application was effective for developing tsunami warning system.

Keywords: Tsunami database, Numerical simulation, Web application.

INTRODUCTION

The occurrence of the Great Sumatra earthquake followed by a devastating tsunami which struck the Indian Ocean region killing more than 230,000 peoples, has renewed Indonesian awareness of the importance of tsunami countermeasures. In order to protect people from tsunami disaster, the Indonesian government launched a project of the Indonesian Tsunami Early Warning System (InaTEWS). The primary task of InaTEWS is to issue reliable tsunami warnings with minimum false-alarm. In order to support it, the development of a pre-calculated tsunami is very significant in issuance of the tsunami warning. According to it, the main objective of this study is to construct a model of tsunami database. Secondary objective is to study the procedure of interpolation techniques using retrieving methods from the database and its application in study area considering the occurrence of tsunami in the future.

---

[*] Meteorological and Geophysical Agency (BMG), Indonesia

[**] Japan Meteorological Agency (JMA), Japan

[***]International Institute of Seismologi and Earthquake Engineering (IISEE), Building Research Institute (BRI), Japan

THEORY AND METHODOLOGY

Tsunami Numerical Simulation

The main purpose of tsunami simulation is to calculate the tsunami heights and its arrival times. Tsunami are commonly referred as long waves, therefore, they are usually modeled mathematically using a depth-averaged approximation of the Navier-Stokes equations referred as "shallow-water wave theory". In this study I use the shallow water theory with Coriolis effect and bottom friction omitted, which are described by Satake (1995). This theory will be solved numerically by applying the leap-frog staggered finite difference method (Satake, 2002). I used the TUNAMI-N2 numerical simulation software which was developed by Disaster Control Research Center of Tohoku University, Japan (Imamura, 2006) to obtain tsunami heights at the target points (forecast and coastal points). In this study, the dimension of computation area was set as 901 grid points and 601 grid points for longitude and latitude direction, respectively. This computation area covered study area from 5.0°S-15.0°S and 102.0°E-117.0°E. The time step of 0.5 s is set to satisfy the stability (CFL) condition. The total calculation time was set to three hours for the completion of the simulation at all the coastal points. To obtain tsunami arrival times I used Tsunami Travel Times (TTT) software which was originally developed by Dr. Paul Wessel from Geoware (http://www.geoware-online.com). In this study I first estimate tsunami heights at forecast points by tsunami simulation, then estimate tsunami heights at coastal point by applying the Green's law to them.

Bathymetry Data

In this study, I use bathymetry data from the General Bathymetric Chart of the Oceans (GEBCO) with resolution of 1 arc minute that is almost equal to 1850 m. The study area is shown in Figure 1 with source points (SP), forecast points (FP) and coastal points (CP).

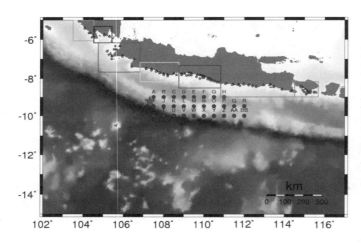

Figure 1. Bathymetry data of study area. Red dots show the source points (30 points) and green dots show forecast points (56 points) and red triangles show coastal points (56 points). Star shows epicenter of the 2006 Pangandaran tsunami. Each small colored rectangle (light blue, blue, magenta, yellow, red, orange and green) shows the area indicating coastal block area.

Source Points

The source point is defined as a center of the fault that could generate tsunami. In this study, a real case source point and artificial ones are used. For the real case, I use the 2006 Pangandaran tsunami (9.222°S, 107.320°E, M_w = 7.7 at 8:19:28 UTC according to USGS). According to the Global CMT solution, the focal mechanism of main shock of earthquake shows strike = 289°, dip angle = 10° and slip angle = 95°. According to Fujii and Satake (2006), the source area is divided into 10 subfaults that cover the aftershock area with the depth of 3 km in the shallow part and 11.7 km in the deep part. All the subfaults have same parameters of focal mechanism with the mainshock.

For the artificial cases, I select 28 source points in the Indian Ocean in front of Java island. For each source point, I use single segment which has the same assumption for fault parameter with strike of 285°. For dip and slip angle, I considered the severest case for tsunami generation and set them to 45° and 90°, respectively. By using these data, I set the depths into 4 scenarios (0, 20, 40 and 60 km) and magnitude into 3 scenarios (7, 7.5 and 8). Then, tsunami simulations were carried out for 337 cases in total including 336 artificial cases and 1 real case.

RESULT OF TSUNAMI SIMULATION

Based on the tsunami simulation results of 336 artificial cases and 1 real case using TUNAMI-N2 software, I will discuss several cases for the sample analysis.

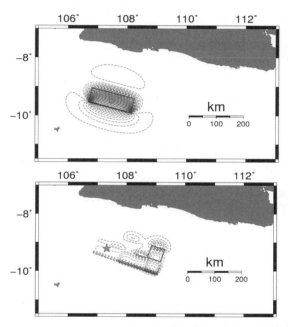

Figure 2. Comparison of fault deformation areas between the artificial case of source point I (top) and the real case of the 2006 Pangandaran tsunami (bottom).

Comparison of tsunami waveforms at coastal area obtained from the simulation between the real case and the nearest artificial source points shows that the tsunami heights of the real case mostly equal to the artificial ones, and arrival times are mostly earlier especially on the central Java coasts. These phenomena appear to be due to the difference of tsunami source area between two cases. Tsunami source area of the real case mostly extends to south-eastern part from the epicenter (main shock) of the real case closely to the central Java coasts like shown in Figure 2. Figure 2 also shows the subsided area of initial sea level displacement of the real case is huge in north-eastern direction, while the subsided area of point I is small for the same direction. This explanation is clarifying why the first arrivals of tsunami for the real case in the central Java coasts mostly subsided while the artificial one is uplifted. Even with those differences in waveforms between the real case and the artificial case, there seems no fatal discrepancy in tsunami heights, but more investigation is needed for arrival times. That indicates validity to use simple rectangular fault model for the artificial cases instead of using realistic but complicated fault model to obtain tsunami heights.

Figure 3 shows the comparison of the tsunami heights at coastal points which were obtained by applying Green's Law to the tsunami heights at different depth of forecast points. Based on this graph, we can find that the tsunami heights obtained at coastal point by applying Green's law are almost the same or do not vary significantly among different bathymetry depths of forecast point. Therefore, considering the complexity of coastal bathymetry along the study area, I decided to use the sea depth of 50 m for forecast points in the tsunami simulations.

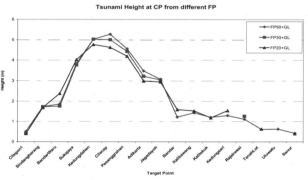

Figure 3. Comparison of tsunami heights at coastal points obtained by applying Green's law to the heights at the forecast points with different depths (source point C with depth of 0 km and magnitude of 8.0).

Figure 4 shows the comparison of tsunami heights from the source point R. From this figure we can see that the application of the Green's law gives relatively higher tsunami heights at coasts than those directly obtained from tsunami simulation. But these results are relatively more stable than the results obtained directly from tsunami simulation. This may be because of not so accurate sea depth data of GEBCO near coasts and choose grid of 1 arc minute, since the complexity of bathymetry of the shallow coastal area could considerably affect the tsunami simulation. Although this result needs further investigation, it gives us good judgement for application of Green's law in the tsunami warning scope.

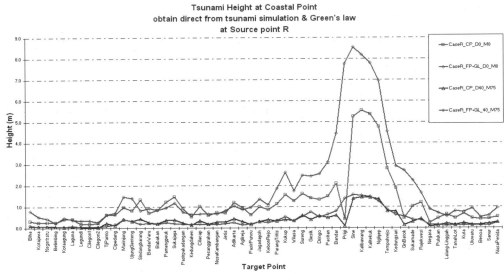

Figure 4. Comparison of tsunami heights obtained directly from simulation and by applying Green's law to heights at FPs for source point R with the same depth and magnitude.

TSUNAMI DATABASE AND WARNING INFORMATION

For establishment of reliable and robust tsunami database, the selection of the database and systems will be important. In this study I adopted a system for developing web application to construct tsunami database and applied retrieve method to it. In this study I used Apache 2.2.8 for web server and MySQL5.0.51 for relational database. PHP 5.2.5 was used in this study as middleware scripting language to connect the server and the database.

Construction of Database

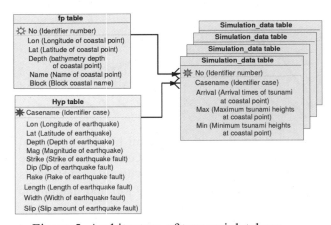

In this study, the tsunami database which I constructed consisted of 3 main parts: 1 coastal point table (fp table), 1 hypocenter table (hyp table) and 336 simulation data tables. I also constructed several temporary tables considering their applications to retrieve from the database. The architecture and relationship between tables of this database are shown in Figure 5. SQL commands were used to make table structure and to insert the tsunami simulation results into the tables.

Figure 5. Architecture of tsunami database.

Retrieving from Database

The next important step to perform database method is to explore how quickly and precisely retrieve tsunami heights and arrival times from the database. In this study I used 2 methods namely simple and interpolation method. These methods are performed by a combination of PHP scripting language and SQL syntax commands. PHP scripting language chooses the most appropriate data from database which match the real earthquake. This combination also makes earthquake and tsunami information message.

Simple method seeks simulation data whose epicenter is located in the nearest horizontal distance to the real earthquake. As for magnitude and depth, this method uses the closest element for outputs. For arrival times, considering the tsunami warning, this method will seek the arrival times from the nearest element to the hypocenter.

To apply the interpolation method, firstly the four cases surrounding the epicenter of the real earthquake will be searched in the database. Each case has 2 different magnitudes and 2 different depths. Totally there exist 16 cases used for the interpolation. In order to obtain tsunami heights for the actual hypocenter, the linear method will be used in horizontal interpolation for 4 cases surrounding the hypocenter. I performed logarithm method to obtain coefficients for magnitude and depth interpolations following the JMA's rules. For estimation of arrival times of tsunami at coastal area, the method will seek the earliest case among the data which were used for the interpolation.

Tsunami Warning Information

Based on JMA (2007), in this study the tsunami warning messages are divided into 3 stages of warning. *The first warning* message consists of data about the recent earthquake that occurred and the earliest arrival time and the highest tsunami height at each estimated coastal block which will be affected. Besides this the information in the warning also includes criteria of warning for each coastal block and name list of the coastal area affected within coastal blocks. *The second warning* message clarifies the details of arrival time and tsunami height at each coastal area within coastal block which were affected. *The third warning* message is about cancellation of tsunami warning information or downgrading or upgrading on the coastal blocks warning criteria. In this study, tsunami warning message was developed by utilizing PHP scripting language integrated with the application of retrieving method in the web application framework. The flowchart of web application used in this study is shown in Figure 6.

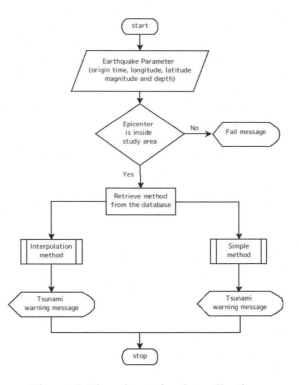

Figure 6. Flowchart of web application.

CONCLUSIONS

In this study, a prototype of tsunami database developed and retrieving methods were used for web application. The source points in the study area are combinations of 1 real case (The 2006 Pangandaran tsunami) and 28 artificial cases with 4 different depths and 3 different magnitudes (totally 337 cases). I considered tectonics setting of study area and the severest case of tsunami occurrence for initial conditions for numerical tsunami simulation.

The tsunami simulation was conducted by using TUNAMI-N2 numerical simulation software from DCRC and TTT software from Geoware. From these simulations I obtained the tsunami heights and estimated arrival times at target points along the study area which consist of 56 forecast and 56 coastal points. I investigated the exact location of forecast points in the offshore area for applying Green's law, and I decided to locate the forecast points in the 50 m depth of bathymetry in the ocean in front of coastal points.

Tsunami simulation result shows that comparison of tsunami waveforms between the real case and the artificial case closest to it had a strong similarity in tsunami heights, but had differences in arrival times due to the differences of fault deformation area for the two cases. Also I found consistency between the tsunami heights at the coast obtained by applying the Green's law and those obtained directly from tsunami simulation. These results show that the setting of artificial cases is acceptable for constructing tsunami database.

Then, the results of tsunami simulation were stored in the database by utilizing web application system. This system consists of Apache 2.2.8 for web server, MySQL 5.0.51 for relational database and PHP 5.2.5 for middleware scripting language. The tsunami database which was successfully constructed in this study consists of 337 tables, and is divided into 3 parts: coastal point table, hypocenter table and simulation data table. Relation between tables in the database based on the primary key from each table has type of 'one to many'. Here, I also succeeded in applying the retrieving method to search the data from the database. The methods I used are simple and interpolation. These methods are easy to apply, but are not so precise due to the lack of bathymetry data I used. So, the accuracy of tsunami heights and arrival times should be increased by using finer bathymetry data. The next stage in this study was to assemble tsunami warning message. The design of tsunami warning message also must be simple and apprehensible by people. Although, I couldn't finish to completing all the tsunami warning message perfectly like the format I designed, this study verified the ability of web application system for developing tsunami warning system. However, there are still needs for further research and improvement to increase utilizing the web application in tsunami warning system.

ACKNOWLEDGEMENT

I wish to express my great appreciation to Dr. B. Shibazaki as Tsunami Course Leader for his guidance, discussions and suggestions during my study in Japan.

REFERENCES

Fujii, Y., and. Satake K., 2006, Source of the July 2006 West Java tsunami estimated from tide gauge records, *Geophysical Research Letter,* 33, L24317, doi:10.1029/2006GL028049.

Imamura, F., Yalciner, A.C., Ozyurt, G., 2006 Revision, Tsunami Modeling Manual, DCRC (Disaster Control Research Center), Tohuku University.

JMA (Japan Meteorological Agency), 2007 Edition, Draft of Manual on Operation Systems for Tsunami Warning Service.

Satake, K., 1995, Linear and nonlinear computations of the 1992 Nicaragua earthquake tsunami, *Pure and Appl. Geophys.*, **144**, 455-470.

Satake, K., 2002, International Handbook of Earthquake & Engineering Seismology: Tsunamis, Academic Press, 437-451.

Synopsis of Master Papers   *Bulletin of IISEE, 43, 139-144, 2009*

PROTOTYPE DATABASE FOR TSUNAMI EARLY WARNING SYSTEM WITH DATA ASSIMILATION IN MALAYSIA

CHAI Mui Fatt*   **Supervisor :   Yushiro Fujii****
MEE07171

ABSTRACT

In order to create a prototype of tsunami database for tsunami early warning system with data assimilation in Malaysia, we located 16 source points in total around the Andaman Sea with 4 magnitudes (M_w6.5, 7.0, 7.5 and 8.0) and 4 depths (0, 10, 20 and 30 km). The coastal and forecast points are located along the Malaysian coastal area at 1 m and 30, 40, 50 and 60 m of bathymetric contour depth with random interval distance, respectively. In numerical simulation, TUNAMI-N2 (Tohoku University's Numerical Analysis Model for Investigation of Near-field tsunamis, No.2) is used to calculate the tsunami waveforms at the output points. Tsunami arrival times at the coastal points are calculated using inverse tsunami travel time by TTT (Tsunami Travel Time). Tsunami database was constructed by using MySQL database which contains of 256 scenario earthquakes that cover historically most active subduction zone around the Andaman Sea. The nearest surrounding data points of a determined hypocenter can be retrieved from database by interpolation, extrapolation and two maximum risk methods. In tsunami data assimilation, TUNAMI-F1 (Tohoku University's Numerical Analysis Model for Investigation of Far-field tsunamis, No.1) is used to calculate the tsunami waveforms at the buoy station of Malaysia. Green's functions, which are calculated tsunami waveforms from faults assigned 1.0 m of slip, are prepared for 16 model sources. In two inversion tests, the initial conditions are precisely resolved by non-negative least squares method. For inversions of two real cases, the slip is almost resolved at the nearest fault and has shown instability in slip distributions for whose epicenter is located closer and slightly out from model sources, respectively.

Keywords: Numerical Simulation, Tsunami Database, Tsunami Data Assimilation.

INTRODUCTION

The Sumatran mega-thrust earthquake that occurred on 26 December 2004 in Indian Ocean has triggered massive tsunami which devastated along the northwest coastal areas of Peninsula Malaysia. In response to this event, Malaysian government has decided to set up the Malaysian National Tsunami Early Warning System in 2005. The purpose of this study is to create a prototype of tsunami database for tsunami early warning system with data assimilation in Malaysia.

THEORY AND METHODOLOGY

Bathymetry Data

In this study, we used GEBCO (The General Bathymetric Chart of the Oceans) with spatial grid size of one arc-minute (~1850 m) to calculate the tsunami travel times and waveforms. The map of bathymetric data in the region of study area is shown in Figure 1.

---

*Malaysian Meteorological Department (MMD), Malaysia

**International Institute of Seismology and Earthquake Engineering (IISEE), Building Research Institute (BRI), Tsukuba, Japan

Tsunamigenic Earthquake Locations

The tsunamigenic earthquake events were searched through Global Centriod Moment Tensor (CMT) Project catalog search from the year 1976 until 2007 (http://www.globalcmt.org/CMTsearch.html). The magnitude and depth range between 6.3 to 10 and 0 to 100 km, respectively. Epicenters chosen lie within 3^0N to 12^0N and 90^0E to 103^0E in latitude and longitude, respectively. Comparison is made with other searcher tsunami databases from Integrated Tsunami Database for the World Ocean (WinITDB) and National Geophysical Data Center (NGDC) Tsunami Event Database (http://www.ngdc.noaa.gov) which covers wider data for a year -2000 to 2007. The comparison has shown that the locations of the tsunamigenic earthquake were located along the subduction zone and Sumatra Fault line (Figure 1).

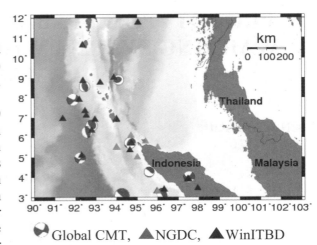

⊕ Global CMT, ▲NGDC, ▲WinITBD

Figure 1. Locations of the tsunamigenic earthquakes with bathymetry data (GEBCO, one arc-minute) in this study area.

Magnitude and Depth

We chosen 4 magnitudes (M_w6.5, 7.0, 7.5 and 8.0) and 4 depths (0, 10, 20 and 30 km), based on the historical earthquake events from WinITDB and Global CMT Project catalog search at study source area. The minimum magnitude of M_w6.5 was chosen based on the tsunami warning criteria of Malaysian Meteorological Department when distant tsunami more than 200 km from Malaysian coastline exists (e.g. Saw, 2007).

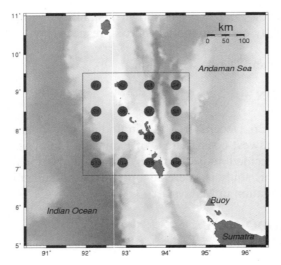

Figure 2. Locations of the 16 source points at Andaman Sea (red circles). The purple triangle shows the location of Malaysian deep ocean buoy.

Source Points

We located 16 source points in total at Andaman Sea northern part of Sumatra which covers the region from 6.83^0N to 9.50^0N and 91.92^0E to 94.58^0E in latitude and longitude, respectively (red line in Figure 2). Each source point is located on the grid with distance interval of 40 arc-minute (~74 km).

Forecast and Coastal Points

17 coastal points and 44 forecast points in total are located along the bathymetric contour depths of 1 m and 30, 40, 50 and 60 m, respectively, with random interval distance. Each bathymetric contour depth of the forecast points is consists of 11 points (Figure 3). The location of the coastal points are placed and searched through Google Earth (2008) considering the most valuable areas for tsunami impacts, denser population areas and tourism attractions. The tsunami heights at the coastal points are estimated by Green's Law.

Green's Law

The Green's Law, conservation of potential energy along the rays (e.g. Satake, 2008), is applied to estimate reliable tsunami heights for coastal points from the forecast points at different bathymetric contour depths. This law is only applicable to direct waves and is not taken account for the reflected waves or edge waves. The tsunami wave front at a forecast point is assumed to be parallel with the one at a coastal point. The tsunami height at a coastal point is calculated by the following equation:

$$h_0 = \left(\frac{d_1}{d_0}\right)^{\frac{1}{4}} h_1$$

Here, h_0 and h_1 are tsunami height at the coastal and forecast point and d_0 and d_1 are water depth at the coastal and forecast point, respectively.

Initial Condition

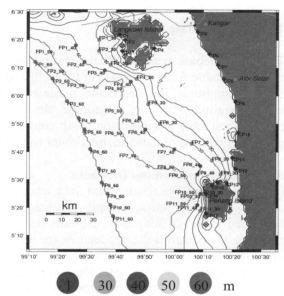

Figure 3. Bathymetry contour map of coastal points (blue dots) and forecast points (green, red, yellow and purple dots).

An initial profile of tsunami source is assumed to be the same as a deformation of ocean bottom due to earthquake when the wavelength of the ocean bottom is much larger than the water depth (Kajiura, 1963). We used the elastic theory of Okada (1985) to calculate the crustal deformation at the ocean bottom due to a fault motion. In this study, a single segment is applied to all 16 source points. For each source point, we assume that the fault model has the same angle of strike (ϕ) as 340^0 (Fujii and Satake, 2007) which is parallel to the trench axis, dip angle (δ) is 45^0 and rake angle (λ) is 90^0. Other parameters such as slip amount (U) in cm, length (L) and width (W) in km which are controlled by moment magnitude (M_w), are determined by Scaling Law (Tatehata, 1997). The equations of Scaling Law are expressed as follows:

$$\log L = 0.5 M_W - 1.9, \quad W = \frac{L}{2}, \quad \log U = 0.5 M_W - 1.4$$

Tsunami Travel Time (TTT)

TTT can calculate tsunami travel times on all of the grid points from a supplied bathymetric data using Huygen's principle (e.g. Fujii, 2008). In this study, tsunami travel times were inversely calculated from the coastal points to the source points. The minimum value of tsunami travel time from a coastal point to the grid points of a deformation source area which has absolute value more than 0.04 m is selected as the tsunami arrival time.

Numerical Simulation by Using TUNAMI-N2

TUNAMI-N2 is applied to shallow water theory in shallow and deep seas. The propagation of tsunami which initiated at each fault is numerically solved by using the finite-difference method (Imamura, 1995). The procedures to run the tsunami numerical simulation are described by Fujii (2008). The dimension of calculation area is 781 and 541 grid points for longitude and latitude, respectively, which covers the region from 90^0E to 103^0E in longitude and from 3^0N to 12^0N in latitude. The temporal interval (Δt) is 3 s to satisfy the CFL (Courant Friedrics Lewy) stability condition. The calculation time was set to 12 hours. The total number of computations is 256 cases for the source points.

Tsunami Database

Tsunami database was constructed by using MySQL database which consists of 258 tables, out of which one table for "FP" and "HYP" each, and 256 tables for simulation results. FP table contains the number, longitude, latitude, depth, name and block of the coastal points. HYP table contains the case name, longitude, latitude, magnitude, depth, strike, dip, rake, length, width and slip of the source points. Each table in simulation result contains the number of the coastal points, case name of source point, arrival time of tsunami, maximum tsunami height and its time at coastal points.

Retrieving from Tsunami Database

Input data to retrieve results from the tsunami database are recently determined hypocenter parameters such as longitude, latitude, depth and magnitude. Four corners surrounding the input data point are contained values of longitude, latitude, depths and magnitudes. The tsunami heights at each nearest corner with a determined hypocenter are retrieved from the database. The tsunami heights according the determined hypocenter are retrieved by using epicenter location, magnitude and depth interpolation methods. Otherwise, the extrapolation method is performed when no surroundings data point is available. In maximum risk method 1 or 2, data point which gave the maximum tsunami height at each coastal point among the source points within the circle area with a half of fault length or the rectangular fault area, respectively, is selected as the database output. We assumed that the recently determined hypocenter is located at 93.250^0E and 8.167^0N in longitude and latitude with $M_w8.5$ and 12 km depth, respectively.

Tsunami Data Assimilation

Initial Condition

The deformation on the ocean bottom is computed for each fault with 1.0 m of slip by the equations of Okada (1985). This displacement is used as an initial condition for the synthetic waveform or Green's function from each fault.

Numerical Simulation by using TUNAMI-F1

TUNAMI-F1 is applied to linear theory for tsunami propagation over the ocean in the spherical coordinates system and numerically solves the governing equations by using finite-difference method (e.g. Nagano, 1991). The area of the numerical calculations covers the region from 90^0E to 103^0E in longitude and from 1^0N to 14^0N in latitude. The dimension of calculation area is 781 and 781 grid points for longitude and latitude, respectively. A temporal interval of 3 s was used to satisfy the CFL stability condition. The computation time of numerical simulation was set to 12 hours.

Tsunami Waveform Inversion

The observed tsunami waveform at the buoy station is expressed as a linear superposition of the computed waveforms from all the faults. The slip amount on each fault can be estimated by using the inversion of non-negative least squares method (Lawson and Hanson, 1974). In numerical simulation, 16 single faults are selected with $M_w7.5$ and 20 km depth for each source point. The initial conditions of the Green's functions are tested by two cases in which the slips of 1.0 or 2.0 m of are assigned to a faults. For real case inversions, two cases with $M_w7.5$ and 20 km depth for each were selected as tsunami generation. The locations of the epicenters are closer or slightly out from model sources.

RESULT AND DISCUSSION

Based on the numerical simulation and TTT results, the tsunami heights and arrival times depend on the magnitude and depth of the source. The tsunami heights and arrival times at coastal points are increased and become faster as the magnitude getting higher. As the source depth is getting deeper, the tsunami heights and arrival times at the coastal points are slightly higher and faster, respectively than a shallower source.

The comparison of results shows that tsunami heights at coastal points obtained directly from numerical simulations (CP) have lower values than the ones at forecast points by the application of the Green's Law (GL) calculations for most cases (Figure 4). Therefore, the tsunami heights calculated with Green's Law are more appropriate and applicable for tsunami warning.

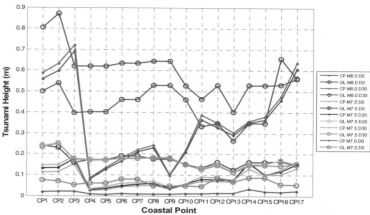

Figure 4. Tsunami heights at the coastal points by different magnitudes and depths.

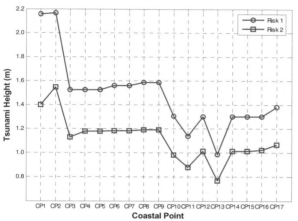

Figure 5. Maximum tsunami heights at the coastal points obtained by the maximum risk method 1 (red line) and 2 (green line).

The database output of maximum risk method 1 has shown that the maximum tsunami height at each coastal point is a combination data from source points of S12 and S8 (red line in Figure 5). For the maximum risk method 2, the source point of S11 has shown the maximum tsunami height at each coastal point and was selected as the database output (green line in Figure 5).

The initial conditions of Test Case 1 and 2 were precisely resolved by the inversion methods and each synthetic waveform is completely agreed with the observed one at the buoy station (Figure 6A and 6B).

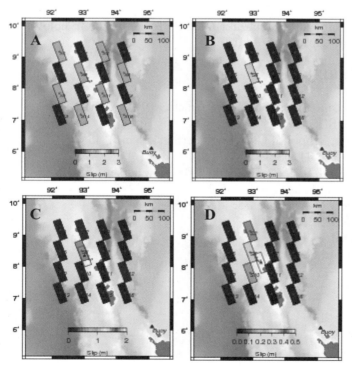

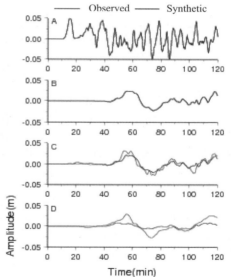

Figure 6. Comparison of observed tsunami waveforms and synthetic ones at the buoy station (right) and inversion results of slip distribution at each fault (left). (A) Case 1, (B) Case 2, (C) Real Case 1 and (D) Real Case 2. The red rectangle is the fault of each real case.

For Real Case 1 (Figure 6C), the largest slip was resolved at fault S6 (1.711 m) which is the vicinity to the epicenter than other faults and the synthetic waveform generally agrees with the observed waveform at the buoy station. However, for the Real Case 2 (Figure 6D) has shown an instability in slip distribution as the largest slip was estimated at fault S2 (0.436 m) and the synthetic waveform is not well reproduced.

CONCLUSIONS

A prototype of tsunami database for tsunami early warning system with data assimilation was successfully constructed. In tsunami database, the nearest data points of a determined hypocenter are retrieved from database by interpolation method. Otherwise, extrapolation method is performed when no surrounding data is available. To optimize the most severe case of tsunami event at each coastal point, two maximum risk methods can be applied.

According to the result of analysis, variation of tsunami heights and tsunami arrival times at coastal points depend on magnitude and depth of the source. The tsunami heights and travel times at coastal points are increased and faster, respectively, as the magnitude becomes larger. On the other hand, the tsunami heights and travel times at the coastal points are slightly higher and faster as the source depth is deeper.

For tsunami data assimilations, the slip amount distributions were determined by using the inversion of non-negative least squares method. The initial conditions of the Green's functions were tested by Case 1 and 2. They were precisely resolved and the synthetic waveform completely agreed with the observed one at the buoy station. For the Real Case 1, the slip distributions are almost resolved at the nearest fault and the synthetic waveform generally agreed with the observed one at the buoy station. However, the Real Case 2 has shown the instability in slip distribution and the synthetic waveform were not well reproduced to match with the observed one at the buoy station.

Combination of the tsunami database by pre-computed scenario earthquakes and the tsunami data assimilation for source estimation will be the useful tool for tsunami forecasting.

AKNOWLEDGEMENT

I would like to express my gratitude to Dr. Bunichiro Shibazaki (IISEE, BRI) for their helpful discussions, valuable comments and guidance during my study in IISEE.

REFERENCES

Fujii, Y., and Satake, K., 2007, Bulletin of the Seismological Society of America, 97, pp. S192-S207.
Fujii, Y., 2008, IISEE Lecture Note 2007-2008, IISEE, BRI.
Imamura, F., 1995, manuscript for TUNAMI code, School of Civil Engineering, Asian, 1-45.
Kajiura, K., 1963, Bulletin Earthquake Research Inst, 41, 535-571.
Lawson, C. L., and R. J. Hanson, 1974, Prentice Hall, Inc., Englewood Cliffs, New Jersey, 340 pp.
Nagano, O., Imamura, F. and Shuto, N., 1991, Natural Hazards 4, 235-255.
Okada, 1985, Bulletin of the Seismological Society of America, 75, 1135-1154.
Satake, K., 2008, IISEE Lecture Note 2007-2008, IISEE, BRI.
Saw, B. L., 2007, IOC, Doc. No. IOC/PTWS-XXII/14.16.
Tatehata, H., 1997, Perspectives on Tsunami Hazard Reduction, 175-188.

NUMERICAL MODELING ANALYSIS FOR TSUNAMI HAZARD ASSESMENT IN WEST COAST OF SOUTHERN THAILAND

Sorot Sawatdiraksa[*]
MEE07174

Supervisor: Fumihiko IMAMURA[**]

ABSTRACT

This study is intended to know the tsunami effect on the west coast of southern Thailand from potential fault locations in and around Andaman Sea. Four cases of fault were simulated as the tsunami source models. TUNAMI code (TUNAMI-N2) used to compute tsunami propagation in the ocean. For large area (region 1), tsunami was computed using GEBCO 1 arc minute bathymetry by the linear theory in spherical coordinate system with Coriolis force. Finer spatial bathymetry grids resample from GEBCO, 15, 5 and 1.67 arcs second in small area, were used in regions 2, 3 and 4 and the tsunami propagation was computed by the non-linear theory in Cartesian coordinate system with bottom friction term. Calculated tsunami heights and travel times were obtained at following output points namely: two existing DART buoys, eight existing tide gauges, four TMD planning tide gauges and forty assumed tide gauges.

The results of Case 1 from linear long wave numerical model have shown that tsunami wave heights were larger than observed data. The results of the non-linear long wave numerical model demonstrated a good agreement with the observed data for tsunami heights and travel times.

The results of numerical simulations have shown that the Thai DART buoy is useful to indicate a generation of tsunami. Tsunami arrived at the Thai DART buoy in Cases 1, 2 and 3 before the other tide gauges. This result can be used to improve accuracy of tsunami warning system for preventing false alarm. Moreover, maximum tsunami heights were larger in the Phangnga and Phuket provinces then integrated countermeasures and city planning must be set up to prevent communities along coastal line.

Keywords: Tsunami simulation, Potential earthquakes, Southern Thailand.

INTRODUCTION

The 2004 Indian Ocean tsunami has made us understand more about the characteristics of tsunami. As the Indian Ocean tsunami occurred in technology age, new information and physical data of tsunami image were observed from satellite altimeters, water level recorded by coastal tide gauge stations around Indian Ocean, tsunami inundation and run-up height measured by field survey including waveforms recorded by many seismographs. However, lack of tsunami early warning system, countermeasures and knowledge caused severe damages along the Indian Ocean coast.

Historical record of tsunami event in Thailand is unclear because there is no evidence of Paleo-Tsunami and effect of tsunami. Some villages have legends related to tsunami events. Morgan people who roam freely from one island to another of the Andaman Sea have been keeping their legend that they must escape to high ground when abnormal sea drag appeared. It is said that Morgan people survived from the Indian Ocean tsunami on 26th December 2004.

---

[*]Thai Meteorological Department, Thailand.
[**]Disaster Control Research Center, Tohoku University, Japan.

METHOD OF COMPUTATION FOR TSUNAMI HAZARD ASSESSMENT

To simulate tsunami propagation in the open sea, we used a linear equation to describe a shallow-water wave on the spherical earth with the Coriolis force, and to simulate tsunami in the coastal zone and the inundation area, we used a non-linear equation in a Cartesian coordinate system for a shallow-water wave with bottom frictions (Goto et al., 2007).

Target Area

We selected wider coarse grid area for far-field tsunami simulation. The position started from latitude 0° N to 25°N and longitude from 75°E to 110°E. This area was called region 1 with bathymetry data from GEBCO (General Bathymetric Chart of the Oceans) and spatial grid size of 1 arc minute. Computation domains in region 1 were 2101 x 1501 grid points and the temporal grid size was set to 1.5 s which satisfied a C.F.L. stability condition and the computation time was 12 hours. Computation domains in region 2 was 825 x 905 grid points which covered area between latitude 5°59'00"N to 9°45'15"N and longitude 96°29'00"E to 99 °55'15"E with a finer spatial grid size of 15 arc seconds.

Region 3 with a finer spatial grid size of 5 arc seconds covered the area between latitude 7°45'25"N to 8°12'39"N and longitude 97°59'55"E to 98 °20'00"E, and had 241 x 325 grid points. Region 4 with the finest spatial grid size of 1.67 arc seconds in this study focused on Kamala and Patong beach in Phuket province and covered the area between latitude 7°50'58"N to 8°00'00"N and longitude 98°5'28"E to 98°19'00"E with 487 x 325 grid points. Non-linear long wave model with bottom friction term was used in region 2, 3, and 4. The coverage areas of region 1, 2, 3 and 4 are shown in Figure 1.

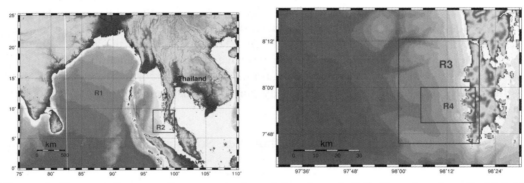

Figure 1. Computation area for numerical model regions 1, 2 (left) and regions 3, 4 (right).

Target Points

We calculated tsunami heights and travel times at 54 stations including the Thai DART buoy station and the Indonesia DART buoy station. Eight tide gauge stations are presented. Within these tide gauges, Taphaonoi and Tarutao tide gauge stations were operated by Hydrographic Department Royal Thai Navy. Ranong, Kuraburi, Krabi, and Kantang tide gauge stations were operated by Marine Department of Thailand. At four tide gauge

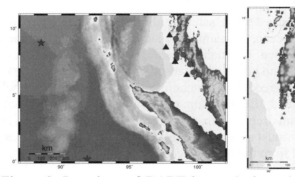

Figure 2. Locations of DART buoys (red stars), existing tide gauge stations (pink circles), and TMD planned installation tide gauge stations (blue triangles). Locations of virtual tide gauge stations (red triangles) along the west coast of southern Thailand.

stations the installation is underway by Thai Meteorological Department. Another tide gauge stations assumed along west coast of southern Thailand near the famous beaches and important residential areas in the provinces. Locations of all tide gauge stations are shown in Figures 2.

Source Model in Andaman Sea

Fault parameters such as fault mechanism, location, and magnitude of earthquake are significant to make generation of tsunami. Tsunami power is related to volume of water change which depends on fault parameters of earthquake. We considered fault locations around Andaman Sea which potentially generate tsunami along Sunda trench and Burma ridge. The source models include 4 cases namely; Case 1 is real tsunami case of the 2004 Sumatra Indian Ocean tsunami in which fault model parameter from DCRC (Disaster Control Research Center, Tohoku University, Japan) were used. Magnitude for this event was estimated to be 9.3 and its fault was divided into 6 sub faults. For second case, we considered height potential seismic gap in north Andaman Sea and the fault area was located from northern part of Andaman Island along Sunda trench. The empirical formula introduced by Papazachos et al. (2004) was used to the estimate fault parameters for an assumed magnitude of 9.0.

The Same earthquake magnitude of 9.0 was used in Case 3. Location of fault along trench line in western side of Andaman Island was similar to that of Fujii and Satake (2007). For Case 4, historical earthquake records were used to provide fault parameters and assumed magnitude was 7.5 in central Andaman Sea. Mechanism of fault was different from the other cases because it was categorized in normal fault. All the fault parameters for four cases are shown in Table 1.

Table 1. Tsunami source parameters for scenario earthquakes in and around Andaman Sea

Parameter	Case 1						Case 2	Case 3	Case 4
Magnitude	9.3						9.0	9.0	7.5
Sub Fault	1	2	3	4	5	6	-	-	-
Length (km)	200	125	180	145	125	380	575.4	575.4	77.6
Width (km)	150	150	150	150	150	150	144.5	144.5	25.1
Displacement (m)	14	12.6	10	11	7	7	9.55	9.55	3.80
Depth (km)	10	10	10	10	10	10	3	3	35
Strike (°)	323	335	340	340	345	7	25	8	254
Dip (°)	15	15	15	15	15	15	10	10	56
Slip (°)	90	90	90	90	90	90	90	115	-90
Latitude (°N)	3.03	4.48	5.51	7.14	8.47	9.63	11.8	8.60	10.95
Longitude (°E)	94.90	93.82	93.30	92.74	92.28	91.97	91.0	91.64	95.05

RESULTS AND DISCUSSION

Scenario Case 1

The first tsunami wave that reached to the Thai DART buoy (by23401) took approximately 27.75 min after the earthquake. The wave was a positive wave with the maximum wave height of 1.055 m. Miang, TMD planning station (tmd007) was the second tide gauge station where the first tsunami wave arrived at approximate time of 62.25 min. Tsunami wave reached Miang station about 34.75 min later than the Thai DART buoy. Simulated tsunami travels time for all the stations are shown in Figure 3.

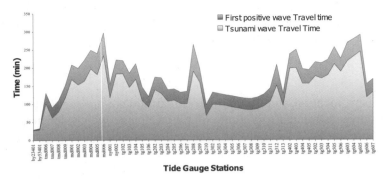

Figure 3. Tsunami travel times and first positive wave travel times (Case 1).

Tsunamis propagate in deep sea faster than in shallow sea. Therefore, waves reached the Thai DART buoy in deep water before they arriving at the other near shore tide gauges.

In this computation case, the maximum tsunami wave height in far-field model was 15.419 m at Kapoe Ranong stations (tg104) and in near-field model was 8.989 m at Khao-lak Phangnga station (tg204).

Tsunami heights and arrival times: comparison between observed data and numerical result

The 2004 Indian Ocean tsunami was observed at six tide gauge stations in Thailand. Tsunami waveforms comparisons between calculated data and observed data at existing tide gauges are shown in Figure 4. The results demonstrated that the tsunami of the far-field model reached to the Ranong, Kuraburi and Kantang stations earlier than the observed data, but the tsunami travel times from calculated data at Taphaonoi, Krabi and Tarutao stations are later than the observed data. For near-field model, earlier tsunami wave reached to the Kuaraburi, Taphanoi, amd Kantang stations before the observed data. However, simulated tsunami reached to the Krabi and Tarutao stations later than the observed data.

Large bathymetry grid size contains one value of the depth, but it will have different values when make a comparison with a finer grid size in the same area. When finer grid is used in one large grid area, the same area provides more detailed values. Tsunami waves which were computed by a non-linear model with a finer grid size and bottom friction terms traveled faster than linear model with larger grid size in a small region. Wave turned to viscosity in large grid size that resulted in smooth waveform shape in far-field model.

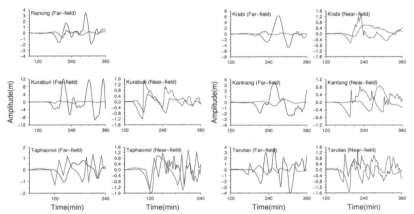

Figure 4. Synthetic waveforms from far-field model (blue colors) and near-field model (green colors) compared with observed tsunami waveforms (red colors) at existing Thai tide gauges for Case 1.

Tsunami heights computed from the far-field model gave over estimated results compared with the observed data. However, computation results have shown a good agreement with observed data in near-field simulation model. For tsunami propagation in the shallow water, where the depth is less than 50 m, bottom friction terms were included in model calculation. Selecting different Manning's roughness value for bottom friction terms can affect the computation result. The maximum tsunami height can be higher or lower than observed data as well as difference of tsunami arrival times. The results of the far-field model have shown that excluding bottom friction terms resulted over estimated tsunami heights compared with the observed data.

Scenario Case 2

Tsunami wave firstly reached to the Thai DART buoy about 38.5 min after the earthquake, which was 56.25 min earlier than the arrival at the second tide gauges, Miang, TMD planning station. The first wave that reached to the Thai DART buoy was a positive wave with the height up to 0.256 m. Tsunami wave reached to the Miang station at approximate time of 94.75 min with the maximum wave height of 0.519 m. Kho-khao Island in Phangnga (tg203) has 2.769 m maximum wave height from far-field model simulation, which was considered as the highest wave in Case 2. The result of the near-field model indicates the tsunami height of 1.311 m at Khao-lak. The maximum wave height distributions concentrate on Indian coastline and other sides as refracted to Myanmar coast.

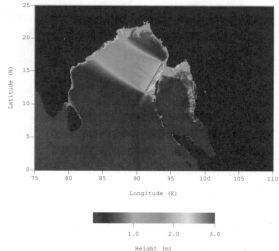

Figure 5. Maximum wave height distribution in Indian Ocean after 12 hours (Case 2).

Tsunami energy was perpendicular to the direction of the fault length. The initial wave was separated into two components and travelled to opposite directions. India is location perpendicular on the major axis of the fault caused the maximum wave height at the coastal line as shown in Figure 5. Tsunami wave can refract in both shallow and deep water. However, the effects of refraction would become prominent near the coast because of the rapid decrease of depth towards the shore. This is one of the reasons why tsunami energy was trapped in shallow water parts of Myanmar.

Scenario Case 3

Similarly to Case 1 and Case 2, tsunami wave reached to the Thai DART buoy firstly at approximate 30.5 min with positive wave and maximum wave height of 0.479 m. The second tsunami wave reached to the Miang station within about 73.75 min. The maximum wave height from the far-field tsunami model appeared at Ban-thap-lamu Phangnga (tg205) with a height of 7.24 m as shown in Figure 6. The

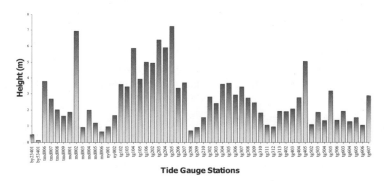

Figure 6. Maximum wave height in Case 3.

result from the near-field model has shown the maximum height of 3.467 m at Kamala beach Phuket (tg305). The maximum wave height distribution shows energy concentration of India coastline and the opposite site, the north of west coast of the southern Thailand.

Scenario Case 4

The simulation results in this case were interesting because of the change of the first station's location. Tsunami wave arrived at the first station at Similan island station (tg210) in about 72 min, which was very close to 72.25 min, the arrival time at the Thai DART buoy. It took about 72.5 min to reach the third station (Miang station) as shown in Figure 7. The maximum wave height distribution is also interesting because tsunami energy was released along large stretches of the coastline in the Andaman Sea. The maximum wave height from the far-field model was 1.632 m at Kamala beach Phuket (tg305) and the wave height from the near-field model was 2.436 m at Naihan beach Phuket (tg309).

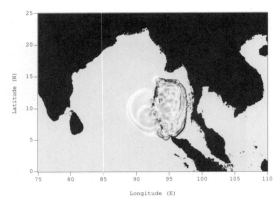

Figure 7. Snapshots of a tsunami wave in Case 4 at time of 72 min after the earthquake.

The tsunami waves in three cases (Case 1, 2, and 3) that reached to the Thai DART buoy at first were positive waves before they arrive at another tide gauges. However, tsunami wave in Case 4 reached to another tide gauge before the Thai DART buoys with negative wave.

The results of the four TMD planning tide gauge stations have indicated that tsunami wave reached Miang Phangnga station earlier than the other stations. The first waves of tsunami that reached to the four TMD planning tide gauge stations were negative in all the cases. The maximum wave height in Case 1 is larger than that in the other cases and the smallest maximum wave height was calculated in Case 4.

CONCLUSIONS

The results of TUNAMI model simulation with the sources around Andaman Sea have shown that the tsunami travel time and wave height in Case 1 gave good agreement with those from near-field model when a comparison was made with observed data. Scenario simulation Cases 1, 2, and 3 presented tsunami waves first reach to the Thai DART buoy before the other tide gauge stations. For TMD planning tide gauge, tsunami first arrived at Miang stations in all the cases.

Even if tsunami cannot be prevented or precisely predicted but damages can be reduce through undertaking various countermeasures and mitigation plans. Moreover, integrated measures are needed to take in to account the achievement of the mission. Adjusting integrated countermeasures must be considered for a suitable area with maximum efficiency. As the results of this study, tsunami heights were larger in Phangnga and Phuket provinces. Therefore, people who live in these areas and along coastal line should be protected from tsunami disaster by the government.

Because the simulated tsunamis reach to the Thai DART buoy at first as mentioned in the result, we can use the Thai DART buoy data as reference in order to assure generation of tsunami for preventing false alarm for tsunami early warning system. However, we must be careful about its use for some cases because in Case 4 the tsunami reached to some tide gauges before the Thai DART buoy.

AKNOWLEDGEMENT

The author would like to express thanks to Dr. Yushiro Fujii, my advisor for reviewing the manuscripts and giving valuable comments.

REFERENCES

Fujii, Y., and K. Satake, 2007, Bulletin of the Seismological Society of America, 97, S192-S207.
Goto, K., S. A. Chavanich, F. Imamura, P. Kunthasap, T. Matsu, K. Minoura, D. Sugawara, and H. Yanagisawa, 2007, Sedimentary Geology 202, 821-837.
Papazachos, B. C., E. M. Scordilis, D. G. Panagiotopoulos, C. B. Papazachos, and G. F. Karakaisis, 2004, Bulletin of the Geological Society of Greece, Vol. XXXVI, Proceedings of the 10th International Congress, Thessaloniki, April 2004.

Technical Information for Contributors

1. The original manuscript will be printed as it is without any changes made. The manuscript, therefore, must be carefully prepared before they are submitted. An Original Paper should not exceed 14 pages including tables and figures. In case of Letter and Technical Notes, 4 pages are allowed. For News and Book Review, 1 page at maximum.

2. Page Format: Use A4 white paper sheets (21 cm x 29.7 cm). Leave 2.5 cm margins at the top, right and left sides of the text and 3.5cm margin at the bottom. Special attention has to be paid in preparing papers using US letter-size paper. It should be appropriately arranged so that it conforms to the above requirements in appearance, namely the manuscript should occupy 16cm x 23.7cm in each page. All main text should be single spaced, Times New-Roman types. Use 18pt in capital letters and boldface for **TITLE**, 12pt for authors, and 11pt for the rest, including affiliations, abstract, main text, headings, sub-headings, sub-subheadings, acknowledgement, appendix, references, and captions for figures, photos and tables.

3. Organization of the papers: Write the **TITLE** of your paper, centered and in 18pt capital letters and boldface types at the top of the first page. After two more line space, write your names in 12pt. Last names should be in capital. Affiliations should be cited by superscripts. Leave two lines, and then write abstract in 11pt. "**ABSTRACT**" should be in capital letters and boldface and be followed by the text of Abstract. After three lines, start main body of your paper in 11pt. The ordinary pages, starting from the second page, contain the main text from the top line. Avoid footnotes and remarks. Explain in the main text, or in Appendices, if necessary. Affiliation itself should be put at the bottom of the first page, cities, countries and e-mail addresses of all authors, as indicated above.

4. HEADINGS: Use at most three levels of headings, i.e., headings, subheadings and sub-subheadings. Headings shall be written in capital letters, boldface types, and centered of your text. Leave two lines space before headings and one after them. Do not indent the first line after headings, subheadings and sub-subheadings. First lines of the other text paragraphs should be indented as indicated here. Do not leave blank lines between paragraphs. **Subheadings:** Subheadings shall be written in lower-case letters and boldface types, right against the left side of your text, as indicated here. Leave one line space before and after subheadings. Use the above mentioned rules for indentation. *Sub-subheadings:* The only difference with respect to subheadings is that sub-subheadings shall be in Italic and no lines space shall be left after sub-subheadings. Don't put numbering to heading of any level.

5. EQUATIONS AND SYMBOLS: Use high quality fonts for both mathematical equations and symbols. Papers with hand-written mathematical equations and symbols are not accepted. Equations should be centered and numbered. Leave one line above and below equations. The equation number, enclosed in parentheses, is placed flush right. Equations should be cited in the text as Eq. (1).

6. FIGURES, TABLES AND PHOTOS: Figures and tables shall be legible and well reproducible, and photos shall be clear. Colored figures, tables and photo will be printed in Black and White. Captions shall be written directly beneath figures and photos and above tables, and shall be numbered and cited as Figure 1, Table 1 or Photo 1. They should be written in 11pt, and centered. Long captions shall be indented. Do not use capital letter or boldface types for captions. Figures, tables and photos shall be set possibly close to the positions where they are cited. Do not place figures, tables and photos altogether at the end of manuscripts. Figures, tables and photos should occupy the whole width of a page, and do not place any text besides figures, tables and photos. Leave one line spacing above and bottom of figures, tables and photos. Do not use small characters in figures and tables. Their typing size should be at least 9pt or larger.

7. UNIT: Use SI unit in the entire text, figures, and tables. If other units are used, provide it in parentheses after the SI unit as 2MPa (19.6 kg/cm^2).

8. CONCLUSIONS: Write a **CONCLUSIONS** section at the end of your paper, followed by ACKNOWLEDGEMENT, APPENDICES and REFERENCES.

9. ACKNOWLEDGMENT: Acknowledgment should follow CONCLUSIONS.

10. APPENDIX: Appendix should be placed between Acknowledgment and References, if any.

11. REFERENCE: All references should be listed in alphabetical order of the first author's family name. They are referred in the main text like (Gibson 1995a). Write the reference list as

Gutenberg, B., and Richter, C. F., 1954, Seismicity of the Earth and Associated Phenomena, 2nd ed. Princeton Univ. Press, Princeton, NJ.

Richter, C. F., 1935, An instrument earthquake magnitude scale, *Bull. Seis. Soc. Am.* **25**, 1-32.

12. Date of acceptance: This will be assigned after accepted for publication and added to the end of manuscript by Editorial Board. They should be written in parentheses in 9pt in boldface types.

13. Download: the template file that may make your editing task easier from http://iisee.kenken.go.jp/ .

CORRESPONDENCE

Manuscript and correspondence should be addressed to:

Editorial Office,
International Institute of Seismology and Earthquake Engineering,
Building Research Institute,
1-Tachihara, Tsukuba, 305-0802, JAPAN
Tel: +81-29-879-0679, Fax: +81-29-864-6777, e-mail:iisee@kenken.go.jp

Electronic submission via e-mail is strongly encouraged to simplify the editing process. The file should be readable and modifiable by MicroSoft Word or by Adobe Acrobat.